ESSENTIAL MILITARIA

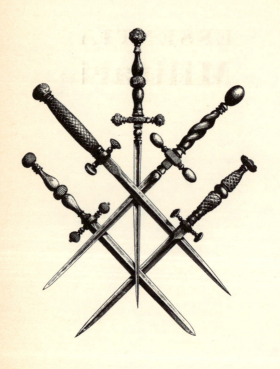

NICHOLAS HOBBES

ESSENTIAL
Militaria

Atlantic Books

LONDON

Published in Great Britain in 2003
by Atlantic Books, an imprint of Grove Atlantic Ltd.

The author and publisher wish to thank the following
for permission to quote from copyrighted material:
George Sassoon for 'The General' Copyright © Siegfried Sassoon;
A. M. Heath & Co. Ltd on behalf of Joseph Heller and
Random House UK Ltd for an excerpt from
Catch-22, Copyright © Joseph Heller 1955

ISBN 1 84354 229 3

Atlantic Books
An imprint of Grove Atlantic Ltd
Ormond House
26–27 Boswell Street
London WC1N 3JZ

Poor Reasons for War

1. FRENCH PASTRIES: In 1838, a squadron of French warships was sent to extract compensation from the fledgling nation of Mexico for French business losses during recent rebellions. One claim was for pastries taken from a restaurant by Mexican leader Santa Anna's troops. The resulting conflict was known as the War of the Cakes.

2. A DETACHED EAR: The War of Jenkins' Ear was named after Robert Jenkins, a British sea captain, who claimed that Spanish coastguards had cut off his ear in 1731. His exhibition of the ear in the House of Commons inflamed public opinion against the Spanish and war was declared in 1739.

3. LOSING A SOCCER MATCH: In 1969 a border dispute turned into a war when El Salvador's football team scored a last-minute winner against neighbouring Honduras in a 1969 World Cup play-off. Hostilities began within hours and the Soccer War was to leave 3,000 dead and 6,000 wounded.

4. SELLING NEWSPAPERS: Press magnate William Randolph Hearst gave his *New York Journal* the edge in a circulation war with the rival *World* by launching a successful campaign for US intervention in the Cuban struggle for independence, leading to the Spanish–American War of 1898.

5. POSTAGE STAMPS: When Bolivia issued a postage stamp featuring a map of its territory that included the disputed border region of Gran Chaco, Paraguay responded by issuing a larger stamp, including Chaco in its own map. The stamps became bigger and bigger until the two sides came to blows in 1932.

6. A GREEDY PIG: The Pig War of 1860 almost broke out on the US–Canadian border, which a Canadian pig kept crossing in order to eat American potatoes. When the American farmer shot the beast, a British warship was dispatched to San Juan resulting in a stand-off with sixty US soldiers. Fortunately, the commanders of each side agreed to stand down.

7. BULL ENVY: According to the Irish legend of the War of the Brown Bull, Queen Medb of Connaught was jealous of her husband's great bull Finnbhennach ('white-horned'), so she led armies from four provinces to capture the bull Donn from an Ulster chieftain.

Delicacies of War

Eaten by the *Daily News* Correspondent, Henry Labouchère, during the Siege of Paris by the Germans in 1870

1. Roast cat (tasted 'like squirrel… delicious')
2. Kittens in onion ragout ('excellent')
3. Donkey steaks ('like mutton')
4. Rat salami ('something between frog and rabbit')
5. Spaniel slices ('by no means bad, something like lamb')

The Seven Past Lives of General George S. Patton

1. A prehistoric mammoth hunter
2. A Greek hoplite who fought the Persians
3. A soldier of Alexander the Great at the siege of Tyre
4. Hannibal
5. A Roman legionary under Julius Caesar
6. An English knight during the Hundred Years' War
7. A Napoleonic marshal

Disciplinarians

1. SHAKA: The Zulu king so disliked slow marchers that he would spear the last man in every column on the way to a battle.

2. GENERAL GEORGE S. PATTON: Slapped and punched shell-shocked soldiers and threatened them with his pistol, telling one artilleryman, 'Hell, you are just a goddamned coward, you yellow son of a bitch. Shut up that goddamned crying. I won't have these brave men here who have been shot seeing a yellow bastard sitting here crying. You're a disgrace to the Army and you're going back to the front to fight, although that's too good for you. You ought to be lined up against a wall and shot. In fact, I ought to shoot you myself right now, God damn you!'

3. DUKE OF WELLINGTON: Was against the abolition of flogging in the British Army, believing it to be necessary and better than putting men in detention where they would be unavailable for duty.

4. LYCURGAS OF SPARTA (*c.* 800 BC): Founder of the Spartan 'Constitution' and, by extension, father of the practice of encouraging gay sex for soldiers as a means of bonding.

5. BRIGIDIER GENERAL ANDREW ATKINSON HUMPHREYS: The leader of a Union division at Gettysburg was a maestro of bad language. Charles Anderson Dana, the Assistant Secretary of War, thought him 'one of the loudest swearers' he had ever known, a man of 'distinguished and brilliant profanity'.

6. XERXES OF PERSIA: According to Herodotus, when the tyrant's first attempt to build a pontoon bridge across the Hellespont in 480 BC was ruined by a storm, he had the waters whipped in punishment.

7. LEON TROTSKY: During the Russian Civil War of 1918–20, Trotsky formed 'Blocking Detachments' behind the front lines to shoot soldiers who attempted to retreat.

8. LAVRENTI BERIA: Under the stewardship of Beria, the Soviet Union's NKVD secret police sentenced 400,000 Russians to service in the penal battalions during the Second World War. These units were used for suicide missions, such as clearing minefields by walking through them en masse.

9. **JOSEPH STALIN:** After the purge of the officer cadre in 1937, three of the Red Army's original marshals and thirteen of fifteen army commanders were dead, 55 per cent of divisional and brigade commanders had been executed, 80 per cent of all colonels and 43 per cent of all other officer ranks.

Gays in the Military

Some commanders reputed to be actively homosexual or bisexual

Achilles
Leonidas of Sparta
Alexander the Great
Julius Caesar
Richard the Lionheart
Saladin
William of Orange
Frederick the Great
Lawrence of Arabia

Nicknames for the US Marines

1. **Soldiers of the Sea:** A phrase first used to describe the British seaborne troops of the seventeenth century.
2. **Leathernecks:** Referring to the leather collars used in the nineteenth century for protecting the neck and keeping Marines' heads erect on parade.
3. **Gyrenes:** A derogatory term bestowed by the US Navy around 1900, formed from a combination of 'GI' and 'Marines'.
4. **Devil Dogs:** Coined by the Germans who fought them in 1918.
5. **Jarheads:** A derogatory term from the Second World War, referring to the image presented by the Marines' high-collared dress uniform.

6. **The President's Own:** Extended from its original reference to the Washington DC Marine band which plays at official functions.
7. **America's 911 Force:** Used because the Marines are the first troops to be called upon in times of crisis (911 being the American emergency telephone number).
8. **Faresta:** Meaning 'Sea Angels' and given to the Marines by Bangladeshi flood victims in 1991.

The US Marine Corps Rifleman's Creed

'This is my rifle. There are many like it, but this one is mine. It is my life. I must master it as I must master my life. Without me, my rifle is useless. Without my rifle, I am useless. I must fire my rifle true. I must shoot straighter than the enemy who is trying to kill me. I must shoot him before he shoots me. I will. My rifle and I know that what counts in war is not the rounds we fire, the noise of our burst, or the smoke we make. We know that it is the hits that count. We will hit.

'My rifle is human, even as I am human, because it is my life. Thus, I will learn it as a brother. I will learn its weaknesses, its strengths, its parts, its accessories, its sights and its barrel. I will keep my rifle clean and ready, even as I am clean and ready. We will become part of each other.

'Before God I swear this creed. My rifle and I are the defenders of my country. We are the masters of our enemy. We are the saviours of my life.

'So be it, until victory is America's and there is no enemy.'

The Agoge: Training Methods of the Ancient Spartans

Undertaken by boys from the age of seven

1. Boys were forced to sleep naked in the middle of winter (and permitted only one layer of clothing during the day).
2. Boys were forbidden to wear shoes on long marches to strengthen the soles of their feet.

3. *Meagre rations*: Boys were encouraged to supplement their diet by stealing extra food, though they were beaten severely if caught in the act.
4. *The scourge*: Boys were regularly whipped by their elders and taught to take pride in the degree of pain they could endure.
5. *The cheese game*: Held annually in front of the altar of Orthia Artemis, Spartan boys had to run the gauntlet of older youths armed with sticks and whips and try, while still conscious, to retrieve wheels of cheese.
6. *Periodic killing sprees in the countryside*: The victims were Helot slaves working the land. Each year 'war' was declared on the Helots to keep them in line.
7. *Pitched battles*: Groups of boys pitted against each other in free-for-alls of unarmed combat.
8. *'Grinding the tree'*: A number of boys formed in a line, with each pressing his shield against the boy in front with as much force as he could muster. The boy at the front of the line was ground against a tree until it toppled – a process often lasting several days and often with fatal results.
9. *The 'Oktonyktia' (eight nights)*: Twelve hundred warriors would march in full pack and armour for four nights, bivouacking during the day. During the next four days and nights they drilled almost continuously, breaking only for short rests. They ate half rations for the first four days, had no food for the next two and no food or water for the last two days.

'The General', by Siegfried Sassoon (1886–1967)

'Good morning! Good morning!' the General said
When we met him last week on the way to the Line.
Now the soldiers he smiled at are most of 'em dead
And we're cursing his staff for incompetent swine.
'He's a cheery old card,' grunted Harry to Jack,
As they slogged up to Arras with rifle and pack,
But he did for them both with his plan of attack.

Leaders represented on Belgian Playing Cards during the Second World War

King of spades	WINSTON CHURCHILL
King of diamonds	FRANKLIN D. ROOSEVELT
King of clubs	CHARLES DE GAULLE
King of hearts	JOSEPH STALIN
Joker	ADOLF HITLER

Iraq's Most Wanted

From the set of playing cards issued to US troops during the 2003 Iraq War

Ace of spades	SADDAM HUSSEIN, President
Ace of hearts	ODAI HUSSEIN, Saddam's son
Ace of diamonds	ABID HAMID MAHMUD, presidential secretary
Ace of clubs	QUSAI HUSSEIN, Saddam's son
Two jokers	One lists Arab titles, the other Iraqi military ranks

The Eight Wounds sustained by Alexander the Great

Cleaver slash to the head
Sword blow to the thigh
Catapult missile in the chest
Arrow passed through the leg
Stone struck the head and neck
Dart pierced the shoulder
Arrow in the ankle
Arrow lodged in the lung

'Hobart's Funnies'

Percy Hobart was a former major general in the British Army who, having fallen foul of the establishment, was not promoted further so had to retire in 1939. He then joined the Home Guard. He was restored to active service by Winston Churchill in 1941 and set to work designing the specialized tanks used during and after the D-Day landings.

1. **Duplex Drive (DD)**: A swimming Sherman tank fitted with airbags and propellers.
2. **The Crab**: A mine-clearing Sherman tank fitted with a rotating drum with chains that flailed the ground.
3. **Petard**: A Churchill tank armed with a heavy mortar for destroying pillboxes and strongpoints.
4. **Crocodile**: A flame-throwing Churchill tank which shot liquid fire through pillbox slits.
5. **Bobbin**: A Churchill tank carrying a giant reel of canvas that it would lay down over sand or soft ground. This 10-foot wide carpet would prevent the tank and subsequent vehicles from becoming bogged down.
6. **Armoured Vehicle Royal Engineer tank (AVRE)**: A tank with a variety of uses, including carrying ramps and a portable bridge.
7. **Fascine**: A Churchill tank carrying a bundle of wooden poles lashed together for filling in craters and ditches.

Lieutenant Colonel Tim Collins, commander of the 1st Battalion of the Royal Irish Regiment, 20 March 2003, Kuwait, near the Iraqi border

'We go to liberate not to conquer. We will not fly our flags in their country. We are entering Iraq to free a people and the only flag which will be flown in that ancient land is their own. Show respect for them.

'There are some who are alive at this moment who will not be alive shortly. Those who do not wish to go on that journey, we will not send. As for the others, I expect you to rock their world. Wipe them out if that is what they choose. But if you are ferocious in battle remember to be magnanimous in victory.

'Iraq is steeped in history. It is the site of the Garden of Eden, of the Great Flood and the birthplace of Abraham. Tread lightly there.

'You will see things that no man could pay to see and you will have to go a long way to find a more decent, generous and upright people than the Iraqis. You will be embarrassed by their hospitality even though they have nothing. Do not treat them as refugees for they are in their own country. Their children will be poor, in years to come they will know that the light of liberation in their lives was brought by you.

'If there are casualties of war then remember that when they woke up and got dressed in the morning they did not plan to die this day. Allow them dignity in death. Bury them properly and mark their graves.

'It is my foremost intention to bring every single one of you out alive but there may be people among us who will not see the end of this campaign. We will put them in their sleeping bags and send them back. There will be no time for sorrow.

'The enemy should be in no doubt that we are his nemesis and that we are bringing about his rightful destruction. There are many regional commanders who have stains on their souls and they are stoking the fires of hell for Saddam. He and his forces will be destroyed by this coalition for what they have done. As they die they will know their deeds have brought them to this place. Show them no pity.

'It is a big step to take another human life. It is not to be done lightly. I know of men who have taken life needlessly in other conflicts, I can assure you they live with the mark of Cain upon them.

'If someone surrenders to you then remember they have that right in international law and ensure that one day they go home to their family.

'The ones who wish to fight, well, we aim to please.

'If you harm the regiment or its history by over-enthusiasm in killing or in cowardice, know it is your family who will suffer. You will be shunned unless your conduct is of the highest for your deeds will follow you down through history. We will bring shame on neither our uniform or our nation . . .

'As for ourselves, let's bring everyone home and leave Iraq a better place for us having been there.

'Our business now is north.'

Animals at War

1. ELEPHANTS: The Carthaginians first used them against Rome at the siege of Agrigentum in 262 BC, but the Romans famously learned to open ranks and simply let the beasts charge through. At the Battle of Panoramus in 251 BC, they were maddened with arrows into stampeding among their own lines, while at Zama in 202 BC they were scared away by trumpets. Elephants were also used extensively in India and in tenth-century China, where they were eventually abandoned as being too vulnerable to massed archery. At the siege of the Qusu in China in AD 446, the attackers made bamboo lions which frightened the city's elephants into trampling their fellow defenders.

2. DOGS: Red Army soldiers strapped bombs to dogs to destroy German tanks in the Second World War. However, the animals identified their own armies' vehicles with food and caused several Russian formations to retreat.

3. BATS: During the Second World War, the USA's 'Project X-Ray' involved strapping miniature napalm charges to thousands of bats and releasing them over Japan. The plan was abandoned after the bats escaped and destroyed an aircraft hangar and a general's car in New Mexico.

4. CAMELS: Afghan mujahidin used kamikaze camels loaded with explosives against the Soviet occupation forces between 1979 and 1989. Camels were used as mobile water tankers during the march of Khalid b. al-Walid's Arab army from Iraq to Syria in 634. First they were forced to drink their fill before their mouths were bound up to prevent them chewing the cud. They were then slaughtered as needed and the water drunk straight from their bellies.

5. RATS: The British Special Operations Executive used dummy rats packed with explosive to disable German munitions factories during the Second World War.
6. DOLPHINS: Both the Russian and US navies have trained dolphins to locate mines.
7. SEA LIONS: Used by the US Navy in the Persian Gulf to keep a lookout for enemy frogmen during the Iraq War of 2003.
8. TICKS: The US Army attempted to use bloodsucking insects to detect hidden enemy troops, but their behaviour proved too difficult to interpret.
9. PIGEONS: The Americans trained these birds to ride in the noses of missiles to guide them towards ships, though they have never been used in combat.
10. MONKEYS: In the *Arthashashtra* (*Treatise on Siegecraft*), the fourth-century BC Brahman Chief Minister Kautilya writes of burning out the defenders of strongholds by using trained monkeys to carry incendiary devices over the fortifications.
11. OXEN: At the siege of Jimo in 279 BC, the defending commander Tian Dan of Qi sent out a hundred oxen dressed in silk costumes to make them look like dragons and with burning straw tied to their tails. The attackers fled.
12. PARROTS: During the First World War, trained parrots were perched on the Eiffel Tower, from where they could give 20 minutes' warning of incoming aircraft. The practice was abandoned when it was discovered that the birds could not discriminate between German and Allied planes.

'The Art of Using Troops', from Sun Tzu's The Art of War

'When ten to the enemy's one, surround him
When five times his strength, attack him
If double his strength, divide him
If equally matched you may engage him
If weaker numerically, be capable of withdrawing
And if in all respects unequal, be capable of eluding him,
for a small force is but booty for one more powerful.'

Designations in the Dutch Resistance during the Nazi Occupation

'Princes'	All who carried out acts of resistance
'Priests'	Harbour watchers and saboteurs
'Barristers'	Communications saboteurs
'Brewers'	Power supply saboteurs
'Painters'	Railway saboteurs

Special Forces by Country

AUSTRALIA	Australian Special Air Service Regiment (SASR)
BELGIUM	Equipes Spécialisées de Reconnaissance (ESR)
CANADA	Joint Task Force-Two
DENMARK	JaegerKorpset (Ranger Corps)
EGYPT	Task Force 777
FRANCE	Régiment parachutiste d'infanterie de marine (RIPMa)
GERMANY	Kommando Spezialkräfte (KSK) (Combat Diver Company); Grenzschutzgruppe (GSG-9) (Border Police Force)
ISRAEL	Sayaret Golani (Golani Reconnaissance Company)
ITALY	Commando Raggruppamento Subacqui ed Incursioni (COMSUBIN)
MEXICO	Force F 'Zorros'
NETHERLANDS	Bizondere Bijstands Eenheid (BBE) (Special Backup Unit)
NORWAY	Marine Jagerne (Marine Hunters)
REPUBLIC OF IRELAND	Sciathan Fianoglach an Airm (Army Rangers Wing)
RUSSIA	Spetsnaz
SOUTH AFRICA	1 Reconnaissance Commando
UK	Special Air Service (SAS), Special Boat Service (SBS)
USA	Delta Force, Navy Seals, Green Berets (US Army Special Forces)

English Civil War Propaganda

A Puritan describes his Cavalier enemies

DESCRIPTION OF
AN ENGLISH ANTIC

With a list of his ridiculous Habits
and apish Gestures

Maids, Where are your hearts become?
Look you what here is!

1. His hat in fashion like a close-stoolpan.
2. Set on the top of his noodle like a coxcombe.
3. Banded with a calve's taile, and a bunch of riband.
4. A feather in his hat, hanging down like a fox taile.
5. Long haire, with ribands tied in it.
6. His face spotted.
7. His beard on the upper lip, compassing his mouth.
8. His chin thrust out, singing as he goes.
9. His band lapping over before.
10. Great band strings, with a ring tied.
11. A long wasted dubblet unbuttoned half way.
12. Little skirts.
13. His sleeves unbuttoned.
14. In one hand a stick, playing with it, in the other side his cloak hanging.
15. His breeches unhooked ready to drop off.
16. His shirt hanging out.
17. His codpiece open tied at the top with a great bunch of riband.
18. His belt about his hips.
19. His sword swapping between his legs like a monkey's taile.
20. Many dozen points at knees.

Military Bureaucracy

Memo from the Alaska Air Command, February 1973

'Due to an administrative error, the original of the attached letter was forwarded to you. A new original has been accomplished and forwarded to AAC/JA (Alaskan Air Command, Judge Advocate office). Please place this carbon copy in your files and destroy the original.'

Officer Fitness Reports

Excerpts, supposedly from Royal Navy and Royal Marines evaluation forms, but widely circulated on email, with varying attributions

1. His men would follow him anywhere, but only out of curiosity.
2. I would not breed from this officer.
3. When she opens her mouth, it seems that this is only to change whichever foot was previously in there.
4. He has carried out each and every one of his duties to his entire satisfaction.
5. This young lady has delusions of adequacy.
6. Since my last report, he has reached rock bottom and has started to dig.
7. She sets low personal standards and then consistently fails to achieve them.
8. This officer should go far, and the sooner he starts, the better.
9. Works well when under constant supervision and cornered like a rat in a trap.
10. This man is depriving a village somewhere of an idiot.

Dambusters Targets 17 May 1943

Dams on the River Ruhr

MOHNE DAM: Capacity of 135 million cubic metres of water; built from limestone masonry; breached.

EDER DAM: Capacity of 202 million cubic metres; built from limestone masonry; breached.

SORPE DAM: Capacity of 72 million cubic metres; built from concrete encased by an earthwork embankment; not breached.

The World's Largest Land Armies

1. China – 1,700,000*
2. India – 1,200,000
3. North Korea – 900,000*
4. South Korea – 560,000*
5. Pakistan – 520,000
6. USA – 475,000
7. Myanmar – 325,000
8. Russia – 320,000
9. Iran – 320,000*
10. Egypt – 310,000

*Includes conscripts

UK – 110,000

Iraq – 360,000 pre-Second Gulf War

The World's Largest Air Forces

By number of combat jets

1. Russia – 3,9961
2. China – 3,5201
3. USA – 2,5981
4. India – 7741
5. Taiwan – 598
6. North Korea – 593
7. Egypt – 583
8. France – 531
9. Ukraine – 521
10. South Korea – 488

UK – 462

The World's Largest Navies

By personnel

1. USA – 369,800
2. China – 230,000
3. Russia – 171,500
4. Taiwan – 68,000
5. France – 62,600
6. South Korea – 60,000
7. India – 53,000
8. Turkey – 51,000
9. Indonesia – 47,000
10. North Korea – 46,000

UK – 36,400

Gods and Goddesses of War

Ares	Ancient Greek
Mars	Roman
Camulus	Celtic
Anuke	Ancient Egyptian
Skanda	Hindu
Wurukatte	Hittite
Huitzilopochtli	Aztec
Nacon	Mayan
Ashure	Assyrian
Tyr	Norse/Germanic
Korrawi	Tamil
Aray	Armenian
Hadur	Hungarian (Hunnic)
Laran	Etruscan
Erra	Sumerian
Hachiman	Japanese
Chun-T'I	Chinese

The World's Largest Military Budgets 2003

1. USA – $334bn
2. Russia – $60bn
3. China – $55bn
4. France – $47bn
5. Japan – $41bn
6. Germany – $39bn
7. UK – $33bn
8. Italy – $20bn
9. Saudi Arabia – $18bn
10. Brazil – $13.4bn
11. South Korea – $13bn
12. India – $12bn
13. Iran – $10bn
14. Australia – $9.3bn
15. Israel – $9bn
16. Spain – $8.6bn
17. Turkey – $8bn
18. Canada – $8bn
19. Netherlands – $7bn
20. Greece – $6bn
21. North Korea – $5bn
22. Singapore – $4.5bn
23. Sweden – $4.4bn
24. Argentina – $4.3bn
25. Egypt – $4bn
26. Mexico – $4bn
27. Poland – $3.5bn
28. Colombia – $3.3bn
29. Norway – $3bn
30. Belgium – $3bn
31. Pakistan – $2.5bn
32. Switzerland – $2.5bn
33. Denmark – $2.5bn
34. Oman – $2.4bn
35. Kuwait – $2bn
36. Algeria – $1.9bn
37. South Africa – $1.8bn
38. Finland – $1.8bn
39. Thailand – $1.7bn
40. Malaysia – $1.7bn
41. Morocco – $1.4bn
42. Libya – $1.3bn
43. Portugal – $1.3bn
44. Czech Republic – $1.2bn
45. Angola – $1.2bn
46. Hungary – $1.1bn
47. Syria – $1bn
48. Peru – $1bn
49. Romania – $1bn
50. Venezuela – $0.9bn

* Iraq pre-Second Gulf War – $1.3bn

US Civil War States

CONFEDERACY

Alabama (seceded 11 January 1861)
Arkansas (6 May 1861)
Florida (10 January 1861)
Georgia (19 January 1861)
Louisiana (26 January 1861)
Mississippi (9 January 1861)
North Carolina (21 May 1861)
South Carolina (20 December 1860)
Tennessee (7 May 1861)
Texas (1 February 1861)
Virginia (17 April 1861)

UNION STATES*

California	New Hampshire
Connecticut	New Jersey
Illinois	New York
Indiana	Ohio
Iowa	Oregon
Kansas	Pennsylvania
Maine	Rhode Island
Massachusetts	Vermont
Michigan	West Virginia
Minnesota	Wisconsin
Missouri†	

* Washington D.C. also allied with Union
† Split loyalties, with a Unionist government in the state capitol and secessionist government-in-exile

NEUTRAL STATES

Delaware*	**Kentucky**	**Maryland***

* Garrisoned by the North to prevent their secession

Items in US Military audit 2000/2001

$45,000 for luxury cruises
$38,000 for lapdancing at strip clubs near military bases
$24,000 for a sofa and armchair
$4,600 for white beach sand and $19,000 worth of decorative
'river rock' at a military base in the Arabian desert
$16,000 for a corporate golf membership
$9,800 for Hallowe'en costumes
$7,373 for closing costs on a home
$3,400 for a Sumo wrestling suit
$1,800 for executive pillows

Developments in Weapons Technology

c. 4000 BC	Chariot
c. 4000 BC	Bronze
c. 3500 BC	Wrought iron
c. 3000 BC	Steel
c. 3000 BC	Galley
c. AD 200	Chainmail armour
c. 300	Stirrup
672	Greek fire
1100	Crossbow
1160	Longbow
c. 1200	Gunpowder*
c. 1250	Rockets
c. 1350	Firearms
1451	Mortar
1592	Armoured warship
1718	Machine gun
1776	Submarine
1776	Sea mine
1797	Parachute
1854	Periscope

* Invented centuries earlier by the Chinese, but not developed as a weapon.

1866	Torpedo
1866	Dynamite
1874	Barbed wire
1903	Aeroplane
1915	Poison gas
1915	Depth charge
1916	Tank
1918	Aircraft carrier*
1918	Sonar
1933	Radar
1937	Helicopter
1939	Jet aircraft
1942	Napalm
1943	Night Vision equipment
1943	Guided missile
1945	A-Bomb
1953	H-Bomb
1955	Nuclear Submarine
1968	Anti-ballistic missile†
1977	Neutron bomb
1982	Stealth aircraft

* In 1918 the Royal Navy launched HMS *Argus*. HMS *Furious* had been adapted for the purpose in 1917.
† The Soviet 'Galosh' system was deployed in 1968 for the defence of Moscow from nuclear attack. Since its warheads were themselves nuclear, it is unknown how much protection they might have afforded the city's populace.

Naval Watches

12 noon to 4:00 p.m.	Afternoon watch
4:00 p.m. to 6:00 p.m.	First dogwatch
6:00 p.m. to 8:00 p.m.	Second dogwatch
8:00 p.m. to midnight	First night watch
Midnight to 4:00 a.m.	Middle or mid watch
4:00 a.m. to 8:00 a.m.	Morning watch
8:00 a.m. to 12 noon	Forenoon watch

'Once More Unto the Breach'

From *Henry V* by William Shakespeare, Act III, Scene I

Once more unto the breach, dear friends, once more;
Or close the wall up with our English dead!
In peace there's nothing so becomes a man
As modest stillness and humility:
But when the blast of war blows in our ears,
Then imitate the action of the tiger;
Stiffen the sinews, summon up the blood,
Disguise fair nature with hard-favour'd rage;
Then lend the eye a terrible aspect;
Let it pry through the portage of the head
Like the brass cannon; let the brow o'erwhelm it
As fearfully as doth a galled rock
O'erhang and jutty his confounded base,
Swill'd with the wild and wasteful ocean.
Now set the teeth and stretch the nostril wide,
Hold hard the breath and bend up every spirit
To his full height! On, on, you noblest English!
Whose blood is fet from fathers of war-proof!
Fathers that, like so many Alexanders,
Have in these parts from morn till even fought,
And sheath'd their swords for lack of argument.
Dishonour not your mothers; now attest
That those whom you call'd fathers did beget you.
Be copy now to men of grosser blood,
And teach them how to war. And you, good yeoman,
Whose limbs were made in England, show us here
The mettle of your pasture; let us swear
That you are worth your breeding; which I doubt not;
For there is none of you so mean and base
That hath not noble lustre in your eyes.
I see you stand like greyhounds in the slips,
Straining upon the start. The game's afoot:
Follow your spirit; and upon this charge
Cry 'God for Harry! England, and Saint George!'

Leonardo Da Vinci's Ideas for Weapons

1. A tank powered by horses or men with hand cranks
2. Submarine
3. One-man battleship
4. A giant cog for sweeping away attackers trying to climb fortifications
5. Lightweight portable bridges
6. Catapult missiles with gunpowder warheads and stabilizing fins
7. A ballista (giant crossbow) 76ft long
8. Breech-loading, water-cooled cannons
9. Anti-personnel cannonballs that shatter upon impact
10. The wheel-lock firing mechanism (used centuries later to improve on the match-lock)
11. A chariot with rotating scythes
12. Double-hulled fighting ships
13. Helicopter

Volunteers for the International Brigade during the Spanish Civil War (1936–39)

1. France – 10,000
2. Germany – 5,000
3. Poland – 4,000
4. Italy – 3,500
5. Britain – 2,500
6. USA – 2,500
7. Belgium – 1,700
8. Czechoslovakia – 1,500
9. Yugoslavia – 1,200
10. Latin America – 1,000
11. Canada – 1,000
12. Hungary – 1,000
13. Scandinavia – 1,000
14. Holland – 600
15. Switzerland – 400

The 'Free World Military Forces'

US-allied troops in the Vietnam War

1. Koreans – 48,869
2. Thais – 11,568
3. Australians – 7,672
4. Canadians – 1,200 (*did not fight under the FWMF banner)
5. New Zealanders – 552
6. Filipinos – 552
7. Taiwanese – 29
8. Spaniards – 10

USA – 5,720,000

German U-Boat Aces of the Second World War

Commander	Ships sunk	Tonnage (tons)
Otto Kretschmer	47	274,386
Wolfgang Lüth	47	225,756
Erich Topp	36	198,658
Günther Prien	30	186,253
Heinrich Liebe	34	185,377
Heinrich Lehmann-Willenbrock	25	179,212
Viktor Schütze	34	174,896
Herbert Schultze	26	169,709
Karl-Friedrich Merten	27	167,271
Joachim Schepke	38	161,340

Fastest Fighter Aircraft of the Second World War

Aircraft	County	Speed (mph/kph)
Messerschmitt Me 163	Germany	596mph (959kph)
Messerschmitt Me 262	Germany	560mph (901kph)
Heinkel He 162A	Germany	522mph (840kph)
North American P-51 Mustang	USA	487mph (784kph)
Focke-Wulf Ta 152H	Germany	472mph (760kph)
Republic P-47N Thunderbolt	USA	467mph (752kph)
Lavochkin La-11	USSR	460mph (740kph)
Vickers Supermarine Spitfire XIV	Britain	448mph (721kph)
Yakovlev Yak-3	USSR	447mph (719kph)
Focke-Wulf Fw 190D	Germany	435mph (700kph)

Fastest Post-war Production Fighter Aircraft

Aircraft	Country	Speed (mph/kph)
Mikoyan-Gurevich MiG-25 Foxbat	USSR	2,110mph (3,390kph) – Mach 3.2
McDonnell Douglas F-15 Eagle	USA	1,650mph (2,655kph) – Mach 2.5
Grumman F-14 Tomcat	USA	1,565mph (2,515kph) – Mach 2.37
Mikoyan-Gurevich MiG-23 Flogger	USSR	1,555mph (2,500kph) – Mach 2.35
Sukhoi Su-27 Flanker	USSR	1,555mph (2,500kph) – Mach 2.35

Mikoyan-Gurevich MiG-29 Fulcrum	USSR	1,520mph (2,445kph) – Mach 2.3
Israel Industries IAI F-21 Kfir	Israel	1,520mph (2,445kph) – Mach 2.3
English Electric F 6 Lightning	Britain	1,500mph (2,414kph) – Mach 2.3
Shenyang J-8 Finback	China	1,450mph (2,340kph) – Mach 2.2
Dassault Mirage 2000	France	1,450mph (2,340kph) – Mach 2.2
Panavia F Mk 3 Tornado	Britain	1,450mph (2,340kph) – Mach 2.2

Abandon Ship!

Scuttled fleets

1. THE CONQUISTADORS: In 1521, Hernando Cortes sailed from Cuba to conquer the Aztec empire of Mexico with a fleet of eleven ships. After landing he built a fort at Veracruz, then scuttled his fleet to prevent desertions and show his men that there could be no turning back.

2. THE POTOMAC ARMADA: John Murray, Earl of Dunmore and the last British governor of Virginia, entered the Potomac in 1776 with seventy-two ships. Anchoring at St George Island to take on provisions, his forces were depleted by smallpox and attacks from revolutionary troops. After two weeks the earl decided to scuttle and burn most of his ships and escape with the remainder.

3. THE BLACK SEA FLEET: In 1854, during the Crimean War, the Russians scuttled their fleet to attempt to block off the harbour entrance to Sebastopol rather than fight the British force at sea.

4. THE HIGH SEAS FLEET: At the end of the First World War, most of the German Imperial Navy was sent to Scapa Flow with skeleton crews as a gesture of good faith while armistice talks

were conducted. Fearing that the British would board and steal the ships, the German commander, Rear Admiral von Reuter, ordered the scuttling of the fleet in June 1919. All but eight ships out of seventy-four were later refloated.

5. THE FRENCH NAVY: After their decision to occupy Vichy France in 1942, the Germans mined the harbour mouth at Toulon to prevent the escape of the rump French fleet to North Africa. Crewmen and Free French supporters scuttled their vessels to prevent their capture.

6. U-BOATS: On 30 April 1945, Germany's Admiral Dönitz issued an order codenamed *Regenbogen* for the scuttling of almost the entire fleet in order to preserve the honour of the Kriegsmarine. The Allies forced him to cancel the order, but the U-Boat commanders in the Western Baltic went ahead anyway, sending 232 submarines to the seabed.

A Short History of Biological Warfare

1. **Poisoned Wells:** In the sixth century BC, the Assyrians poisoned enemy wells with rye ergot, while Solon of Athens used the purgative herb, hellebore (skunk cabbage), to poison the water supply during his siege of Krissa. In the fourth century BC, the Greeks contaminated their enemies' wells with animal corpses, while at the Battle of Tortona, Italy in 1155, Barbarossa used human corpses to pollute the enemy water supply.

2. **Poison Arrows:** In the fifth century BC, Scythian archers dipped their arrows in animal dung so that the wounds they caused would become infected.

3. **The Plague:** In 1346–47, the Muslim Tatar De Mussis caused an epidemic of bubonic plague in Caffa on Russia's Black Sea in Crimea by catapulting infected corpses over the city walls. At the siege of La Calle in 1785, Tunisian troops threw plague-ridden clothing into the city. In 1940, a plague epidemic in China and Manchuria followed reported over-flights by Japanese planes dropping plague-infected fleas.

4. **Manure and corpses:** During the siege of Karlstejn in the Holy

Roman Empire in 1422, dead soldiers and 2,000 cart loads of manure were catapulted over the fortifications.

5. **Vintage leprosy:** In 1485 near Naples, the Spanish supplied the French with wine laced with blood from lepers.

6. **Smallpox:** In 1763, during the French and Indian War, British Colonel Henry Bouquet gave smallpox-infected blankets to the Indians at Fort Pitt, in Pennsylvania, resulting in an epidemic.

7. **Anthrax:** During the First World War, the Germans infected nearly 5,000 mules and horses in Mesopotamia and sent anthrax to Romania to infect sheep being transported to Russia. Four people died in 2001 when the US postal system was used to distribute spores.

Ill-Treatment of Prisoners

1. THE BATTLE OF AEGOSPOTAMI, 405 BC: After Lysander's Spartan fleet had defeated the Athenians, 3,000–4,000 prisoners were executed.

2. THE BATTLE OF CHANGPING, 260 BC: 400,000 survivors from the Chinese state of Zhao's 450,000-strong army surrendered to the Qin general Bo Qi. Every last man was put to death.

3. SPENDIUS AND AUTITARIUS: The commanders of the mercenary revolt against Carthage surrendered with 40,000 starving troops in 238 BC. Spendius was crucified and his men were slaughtered.

4. SPARTACUS: After Marcus Licinius Crassus put down the slave revolt of Spartacus in 71 BC, the Romans crucified 6,000 captured slaves along the Appian Way as a warning to others.

5. THE BATTLE OF BALATHISTA, 1014: The victorious Byzantine emperor Basil II blinded his 15,000 Bulgar prisoners, leaving one man in every hundred with one good eye so that he could lead his comrades home.

6. MERCADIER: Richard the Lionheart's dying wish in 1199 was that the crossbowman who shot him should be spared and given a sum of money. Richard's mercenary captain Mercadier had the man flayed alive and impaled instead.

7. THE BATTLE OF AGINCOURT, 1415: Several thousand French prisoners were allegedly executed on the spot at the orders of the English King Henry V so that he could redeploy the soldiers guarding them.

8. THE BRIDGE OVER THE RIVER KWAI: The Thai–Burma Railway was completed in 1943 by 68,000 Allied POWs and 200,000 Asian slaves. Due to the cruelty of their Japanese overseers, 18,000 POWs and 78,000 Asians died in the process.

9. THE SURVIVORS OF STALINGRAD: Of the 90,000 Germans of Von Paulus' 6th Army who surrendered at the city in 1943, only 5,000 returned from the Soviet Union alive, most having spent ten years in captivity.

10. THE NAZIS: An estimated 3 to 4 million Soviet POWs died in German hands from murder, maltreatment, exposure and starvation.

11. GUANTANAMO BAY: Following the terrorist attack on the US on 11 September 2001 some 700 people suspected of Taliban or al-Qa'ida sympathies, some younger than 16 years old, were taken by the US military to a place beyond the jurisdiction of any of the nations involved and kept incommunicado in cages. In contravention of US and international law they were held indefinitely without charge, trial or sentence. By 2003 several suicide attempts had been made.

12. THE VIET CONG: During the Vietnam War, American POWs were routinely beaten, starved, tortured, politically indoctrinated and sometimes executed by their VC captors.

Military Etymology

Ammunition: From the French *munition* (all war essentials).
Bazooka: From its supposed resemblance to the bazoo or kazoo.
Bomb: From the Greek *bombos* (boom).
Bullet: From the French *boule* (ball) – any projectile (cannon or musket).
Chindit: A corruption of *Chinthé*, the name of a mythical Burmese beast, half-lion and half-dragon.

Commando: From the Portuguese word meaning 'command,' which the Boers in South Africa used to identify local militia units.

Grenade: From the Latin *granatus* (seed-filled).

Guerrilla: From the Spanish for 'little war', referring to resisters of the French occupation during the Peninsular War of 1808–14.

Gun: From the Old Norse *Gunnhildr* (a woman's name).

Missile: From the Latin *missilis* (a thrown or fired weapon).

Partisan: From the French word *partis*, used to mean foraging parties.

Tank: The first examples were sent to the Western Front in containers marked 'Petrograand' for secrecy, the cover story being that these were water tanks being developed for the Russians.

Tattoo: A Thirty Years' War corruption of the 'taps' that were used to plug wine barrels at the end of a drinking session.

Technicals: Pick-up trucks with mounted machine guns were given this name during the war in Somalia in 1991–95 by journalists who hired them for protection while filing claims for 'technical' expenses.

Cold War Doctrines

Truman Doctrine: The commitment to contain the communist threat to the world within its 1948 borders.

Eisenhower Doctrine: The commitment to send US troops to the Middle East to counter any communist threat.

Formosa Doctrine: American resolve to protect Taiwan from Chinese communist aggression.

Massive Retaliation Policy of responding to a Soviet attack on NATO with nuclear weapons.

Nuclear Deterrence: The notion that the possession of nuclear weapons is alone enough to prevent foreign invasion.

Reagan Doctrine: Military aid to anti-communist insurgents in the Third World.

Mutually Assured Destruction (MAD): The excuse for new
weapons development on the grounds that the systems
created by both sides would be able to destroy the enemy
even if one's own country were already destroyed, thereby
preventing a 'first strike' by either side.

Domino Theory: The opinion of US President Lyndon Johnson's
advisors that if South Vietnam fell to the communists, the
same fate would befall other Asian countries and then countries
outside Asia.

Brezhnev Doctrine: Soviet policy of invading Warsaw Pact allies if
they moved too far away from orthodox Soviet Communism,
such as Czechoslovakia in 1968.

Napoleon's Medical Problems (real and exaggerated)

1. Colic: When the pain subsided he sometimes had to sleep
   during battles.
2. Peptic ulcers.
3. Dysuria: Making it painful for him to pass urine.
4. Pituitary dysplasia: Stunted his growth.
5. Oedema of the chest: Excess fluid in the lungs, giving him a
   persistent cough.
6. Fevers.
7. Constipation: From childhood onwards.
8. Chronic gastroenteritis (when not constipated).
9. Prolapsed haemorrhoids (alleged): Eventually making it
   impossible for him to mount his horse at Waterloo.
10. Micro-penis: One-inch long and resembling a grape, an organ,
    supposedly the dictator's, was cut off at autopsy and eventually
    put up for auction by Christie's in 1972, where it failed to
    reach the reserve price.

The 'Black Book' – Writers on the 1940 Gestapo hit list should Great Britain have been occupied by the Nazis

1. Vera Brittain
2. Noel Coward
3. Sigmund Freud (died 23 September 1939)
4. Aldous Huxley
5. J.B. Priestley
6. Bertrand Russell
7. C.P. Snow
8. Steven Spender
9. Lytton Strachey (died 2 January 1932)
10. Rebecca West
11. Virginia Woolf

'The Soldier', by Rupert Brooke (1887–1915)

If I should die, think only this of me:
That there's some corner of a foreign field
That is for ever England. There shall be
In that rich earth a richer dust concealed;
A dust whom England bore, shaped, made aware,
Gave, once, her flowers to love, her ways to roam,
A body of England's, breathing English air,
Washed by the rivers, blessed by the suns of home.

And think, this heart, all evil shed away,
A pulse in the eternal mind, no less
Gives somewhere back the thoughts by England given;
Her sights and sounds; dreams happy as her day;
And laughter, learnt of friends; and gentleness,
In hearts at peace, under an English heaven.

Second World War Fighter and Bomber Production

Aircraft	Country	Units
Ilyushin Il-2 Shturmovik	Russia	42,330
Messerschmitt Bf 109	Germany	35,000
Vickers Supermarine Spitfire	Britain	20,531
Focke-Wulf Fw 190	Germany	20,000
Consolidated B-24 Liberator	USA	18,000
Yakovlev Yak-9	USSR	16,769
Republic P-47 Thunderbolt	USA	15,677
North American P-51 Mustang	USA	15,367
Hawker Hurricane	Britain	14,500
Boeing B-17 Flying Fortress	USA	12,800
Grumman F6F Hellcat	USA	12,275
Mitsubishi Zero	Japan	10,500
Lockheed P-38 Lightning	USA	10,037
Messerschmitt Bf 110	Germany	10,000
North American B-25 Mitchell	USA	9,889
Grumman TBF Avenger	USA	9,836
Grumman F4F Wildcat	USA	7,885
De Havilland Mosquito	Britain	7,781
Avro Lancaster	Britain	7,377
Junkers Ju 88	Germany	15,000

Substandard Kit

1. *Civil War footwear*: Some Confederate army issue shoes were made out of wood and stained paper, so that thousands of soldiers and one lieutenant colonel soon marched barefoot.
2. *British radios*: The British Army's out-of-date Clansman radio system led one officer to remark in 2001 that the next war 'had better be fought in a country with Vodafone coverage'.
3. *Italian tank armour*: When Italy entered the Second World War, her tanks and heavy vehicles were so thinly armoured that they all could be penetrated by the lightest anti-tank rifles then in service.

4. *German overcoats*: The troops invading the Soviet Union in June 1941 were provided with no winter clothing or anti-freeze for their vehicles, as Hitler assumed the war would be over before the cold set in.

5. *Argentine bombs*: 55 per cent of the bombs dropped on the British Task Force during the Falklands War of 1982 failed to explode. Had they detonated, a further six to thirteen ships that were merely damaged might have been sunk and the war lost.

Military and Civilian Honours (British)

Victoria Cross (VC)
George Cross (GC)
Most Noble Order of the Garter (KG)
Most Ancient and Most Noble Order of the Thistle (KT)
Most Illustrious Order of St. Patrick (KP)
Knights Grand Cross in The Most Honourable Order of the Bath (GCB)
Order of Merit (OM)
Baronet's Badge
Knight Grand Commander in The Most Exalted Order of the Star of India (GCSI)
Knights Grand Cross in The Most Distinguished Order of St. Michael and St. George (GCMG)
Knight Grand Commander in The Most Eminent Order of the Indian Empire (GCIE)
The Order of the Crown of India (CI)
Knights Grand Cross in The Royal Victoria Order (GVCO)
Knight (or Dame) Grand Cross in The Most Excellent Order of the British Empire (GBE)
Order of the Companions of Honour (CH)
Knight (or Dame) Commander in The Most Honourable Order of the Bath (KCB or DCB)
Knight Commander in The Most Exalted Order of the Star of India (KCSI)

Knight (or Dame) Commander in The Most Distinguished Order of St. Michael and St. George (KCMG or DCMG)

Knight Commander in The Most Eminent Order of the Indian Empire (KCIE)

Knight Commander in The Royal Victorian Order (KCVO)

Knight (or Dame) Commander in The Most Excellent Order of the British Empire (KBE or DBE)

Knight Bachelor's Badge (KB)

Companion in The Most Honourable Order of the Bath (CB)

Companion in The Most Exalted Order of the Star of India (CSI)

Companion in The Most Distinguished Order of St. Michael and St. George (CMG)

Companion in The Most Eminent Order of the Indian Empire (CIE)

Commander in The Royal Victorian Order (CVO)

Commander in The Most Excellent Order of the British Empire (CBE)

Distinguished Service Order (DSO)

Lieutenant in The Royal Victorian Order (LVO)

Officer in The Most Excellent Order of the British Empire (OBE)

Imperial Service Order (ISO)

Member in The Royal Victorian Order (MVO)

Member in The Most Excellent Order of the British Empire (MBE)

Indian Order of Merit (Military) (IOM)

Conspicuous Gallantry Cross (CGC)

Distinguished Service Cross (DSC)

Military Cross (MC)

Distinguished Flying Cross (DFC)

Air Force Cross (AFC)

Distinguished Conduct Medal (DCM)

Conspicuous Gallantry Medal (CGM)

George Medal (GM)

Distinguished Service Medal (DSM)

Military Medal (MM)

Distinguished Flying Medal (DFM)

Air Force Medal (AFM)

Mentions in Dispatches

Some Military and Civilian Honours (US)

Medal of Honor
Navy Medal of Honor
Distinguished Service Cross
Navy Cross
Army Distinguished Service Medal
Navy Distinguished Service Medal
Coast Guard Distinguished Service Medal
Silver Star
Legion of Merit
Distinguished Flying Cross
Soldier's Medal
Navy and Marine Corps Medal
Bronze Star
Air Medal
Naval Commendation Medal
Army Commendation Medal
Coast Guard Commendation Medal

Bombing Raids – Civilian Casualties

1. DRESDEN, 13–14 February 1945
   Firebombing by RAF and USAAF – 25,000 dead
2. TOKYO, 9–10 March 1945
   Firebombing by USAAF – 83,000 dead
3. HIROSHIMA, 6 August 1945
   Atomic bomb 'Little Boy' dropped by the *Enola Gay* –
   78,000 dead
4. HAMBURG, 24 July – 2 August 1943
   Firebombing by RAF – 44,000 dead
5. NAGASAKI, 9 August 1945
   Atomic bomb 'Fat Man' dropped by *Bock's Car* – 40,000 dead
6. DARMSTADT, 11 September 1944
   Firebombing by RAF and USAAF – 11,000 dead

Total British civilian casualties from air raids over the course of
the Second World War = 60,595

Landmines

Estimated total yet to be defused, per country

Iran – 12m
Egypt – 7.5m (after 11m already cleared)
Angola – 6–9m
Afghanistan – 4–10m
Vietnam – 3.5m
Zimbabwe – 2.5m
Eritrea – 2m
Ethiopia – 2m
Morocco – 2m
Mozambique – 1m
Cambodia – 1m–6m
Croatia – 1m
Bosnia-Herzegovina – 1m
Thailand – 1m

There are no official figures for Iraq, but in Iraqi Kurdistan around one person a day steps on a mine.

The Cost of Buying a Commission

In a British Infantry Regiment of the Line, 1830s

1. Lieutenant-Colonel – £4,500 (net annual pay £114) (current: *c.* £180,000)
2. Major – £3,200 (current: *c.* £130,000)
3. Captain – £1,800 (current: *c.* £75,000)
4. Lieutenant – £700 (current: *c.* £28,000)
5. Ensign – £450 (current: *c.* £18,000)

For command of the 11th Hussars, however, Lord Cardigan paid £40,000 in 1860 (current: *c.* £1,600,000).

Notable Warrior Women

1. BOUDICCA: Queen of the Iceni tribe who rebelled against the Roman invasion of Britain in the first century AD.

2. ZENOBIA: Third-century queen of Palmyra who led her armies to defeat at the hands of Aurelian's Romans. Tried for crimes against the empire, she blamed the counsel of the Greek philosopher Longinus, who was executed in her stead.

3. JOAN OF ARC: Successfully led the French army against the English in the early fifteenth century before her enemies accused her of witchcraft and burnt her at the stake when she was 19.

4. ISABELLA I OF CASTILE (1451–1504): The Iberian queen wore armour and led her army in the field against rebels early in her reign and later against the Moors alongside her husband, Ferdinand of Aragon.

5. AMELIANE DU PUGET: Governor's daughter who led a troop of women who broke the siege at Marseilles in 1524 during the war between the King of France and the Constable de Bourbon. They dug a mined trench known as the Tranchée des Dames that became the Boulevard des Dames.

6. GRAINE NI MAILLE: (Grace O'Malley): Irish princess who commanded a large pirate fleet in the sixteenth century. Queen Elizabeth I of England accepted her territorial claims after the two met in 1593.

7. MARGARET CORBIN: American heroine who fought alongside her husband in the Revolutionary War and was the first woman to receive a pension from the United States government as a disabled soldier.

8. EMILIENNE MOREAU: Fought for the French on the Western Front during the First World War. She killed two snipers in the Battle of Loos and was awarded the Croix de Guerre, the British Red Cross Medal and the St John Ambulance Society Medal. In 1940 she once again fought for her country, earning a second Croix de Guerre.

9. ELAINE MORDEAUX: Led around two hundred French Resistance fighters in a guerrilla attack on Germany's 101st Panzer

Division. Though few of the French survived, three hundred German troops were killed in the assault and around a hundred trucks and tanks disabled. The death of Mordeaux and her comrades contributed to the success of the D-Day landings as it held up the division's advance to the coast.

10. LUDMILLA PAVLICHENKO: Russian sniper credited with killing 309 Germans during the Second World War.

11. HANNA REITSCH: The German test pilot was the only woman ever to be awarded the Iron Cross and Luftwaffe Diamond Clasp.

The Crusades

The wars against heretics and infidels between 1096 and 1291 sanctioned by the Pope, initially to ensure the safety of Christian pilgrims in the Holy Land

FIRST, 1095–1101: Walter the Penniless and Peter the Hermit fail to reach the Holy Land, but subsequent expeditions establish a Christian kingdom in Asia Minor with Godfrey de Bouillon, Bohemund de Tankerville and others taking Antioch in 1098 and Jerusalem in 1099.

SECOND, 1145–47: Ended in disaster under Louis VII of France and Conrad III of Germany, who laid siege to Damascus but were driven off with heavy losses after just four days.

THIRD, 1188–92: After Saladin took Damascus in 1174, Aleppo in 1183 and then Jerusalem in 1187, Frederick Barbarossa of Germany invaded through Asia Minor, while Philip Augustus of France and Richard the Lionheart of England landed at Acre and captured the city. After disagreements among the allies, Richard's army was left to fight alone until a truce was concluded.

FOURTH, 1202–4: The crusaders who set out from Venice never reached Jerusalem, but did become embroiled in Venetian politics before sacking Constantinople in 1204 and establishing a Latin empire.

FIFTH, 1217–19: The crusaders invaded Egypt and stormed Damietta in 1219 but then spent a year arguing over the spoils rather than advancing on Cairo. By the time they struck out, their Saracen adversaries had assembled a strong enough force to defeat them and take Damietta for themselves.

SIXTH, 1228–29: The Emperor Frederick II had vowed to lead a crusade as early as 1215 but had used a variety of excuses to avoid his commitment. He eventually embarked for the Holy Land in 1227, but returned a few days later citing ill health. Pope Gregory IX excommunicated him for this final impiety, but Frederick honoured his vow the following year and in 1229 won Jerusalem, Bethlehem and Nazareth by entreaty.

SEVENTH, 1249–52: St Louis, Louis IX of France, took Damietta but was then himself taken prisoner. He was compelled to relinquish the city and pay a ransom of one million gold bezants.

EIGHTH, 1270: St Louis' second mission was hit by the plague soon after he landed at Carthage and forced to re-embark without its fallen leader.

America's Top Ten Defence Contractors (2002)

1. Lockheed Martin Corp. – $17bn
2. Boeing Co. – $16.6bn
3. Northrop Grumman Corp. – $8.7bn
4. Raytheon Co. – $7bn
5. General Dynamics Corp. – $7bn
6. United Technologies Corp. – $3.6bn
7. Science Applications International Corp. - $2.1bn
8. TRW Inc. – $2bn
9. Health Net, Inc. – $1.7bn
10. L-3 Communications Holdings, Inc. – $1.7bn

Countries Using Child Soldiers

Afghanistan	Pre-teen combatants in Northern Alliance forces.
Angola	Children as young as 10 have been forcibly recruited by UNITA rebel and government forces.
Burma	Orphans under 10 are forcibly conscripted.
Burundi	Some soldiers are under the official age of 18.
Colombia	Up to 15,000 child soldiers fight for the FARC communist rebels.
Ivory Coast	Under-15s fight for rebel groups.
DR Congo	Rebel forces continue to recruit children.
Guinea-Bissau	Boys under 16 can volunteer for military service with the consent of their parents or tutors.
Iraq	Before the Second Gulf War, boys who did not volunteer for weapons training in the Saddam Cubs were denied ration cards and exam results.
Liberia	Various armed militias forcibly recruit children.
Nepal	Teenagers are recruited by Maoist rebels.
Paraguay	Up to 100 conscripts under 18 have died since 1989.
Philippines	Rebel and Islamic groups commandeer fighters as young as 11.
Sudan	Both government and rebel forces conscript teenagers.
Uganda	The government can enlist boys under 18 with parental consent. The rebel Lord's Resistance Army is notorious for abducting pre-teens and using them as guerrillas and sex slaves.

'The Charge of the Light Brigade' by Alfred, Lord Tennyson (1809–92)

Half a league, half a league,
Half a league onward,
All in the valley of Death
Rode the six hundred.
'Forward, the Light Brigade!
Charge for the guns!' he said:
Into the valley of Death
Rode the six hundred.

'Forward, the Light Brigade!'
Was there a man dismay'd?
Not tho' the soldier knew
Some one had blunder'd:
Theirs not to make reply,
Theirs not to reason why,
Theirs but to do and die:
Into the valley of Death
Rode the six hundred.

Cannon to right of them,
Cannon to left of them,
Cannon in front of them
Volley'd and thunder'd;
Storm'd at with shot and shell,
Boldly they rode and well,
Into the jaws of Death,
Into the mouth of Hell
Rode the six hundred.

Flash'd all their sabres bare,
Flash'd as they turn'd in air
Sabring the gunners there,
Charging an army, while
All the world wonder'd:
Plunged in the battery-smoke
Right thro' the line they broke;
Cossack and Russian
Reel'd from the sabre-stroke
Shatter'd and sunder'd.
Then they rode back, but not
Not the six hundred.

Cannon to right of them,
Cannon to left of them,
Cannon behind them
Volley'd and thunder'd;
Storm'd at with shot and shell,
While horse and hero fell,
They that had fought so well
Came thro' the jaws of Death,
Back from the mouth of Hell,
All that was left of them,
Left of six hundred.

When can their glory fade?
O the wild charge they made!
All the world wonder'd.
Honour the charge they made!
Honour the Light Brigade,
Noble six hundred!

Epitaphs

TO THE ANZACS

These heroes that shed their blood and lost their lives; you are now lying in the soil of a friendly country. Therefore rest in peace. There is no difference between Johnnies and Mehmets to us where they lie side by side here in this country of ours. You, the mothers, who sent their sons from far-away countries, wipe away your tears; your sons are now lying in our bosom and are at peace. After having lost their lives on this land they have become our sons as well.

Mustapha Kemal Ataturk's words, written in 1934 and inscribed on the First World War memorial to Australian and New Zealand soldiers killed in 1915 at ANZAC Cove in Gallipoli, Turkey

THE CONFEDERATE EPITAPH

Not for fame or reward, not for place or rank, not lured by ambition or goaded by necessity, but in simple obedience to duty as they understood it, these men suffered all, sacrificed all… and died.

In Arlington National Cemetery

THE KOHIMA EPITAPH

When you go home, tell them of us and say,
For their tomorrow, we gave our today.

Written by John Maxwell Edmonds in 1919, inscribed on the Second World War memorial in Kohima, India

TO THE 300 SPARTANS

Go tell the Spartans, you who pass by,
that here, obedient to their laws, we lie.

By Simonides of Kios, inscribed at the Pass of Thermopylae which they held in 480 BC

TO THE BRITISH PARATROOPERS OF 1944 ON ARNHEM BRIDGE

This is the bridge for which JOHN D. FROST fought
leading his soldiers persistent and brave
went a bridge too far which they tried to save
the bridge is now with his name proudly wrought.

The Gettysburg Address, Abraham Lincoln, 19 November 1863

'Fourscore and seven years ago our fathers brought forth on this continent a new nation, conceived in liberty and dedicated to the proposition that all men are created equal. Now we are engaged in a great civil war, testing whether that nation or any nation so conceived and so dedicated can long endure. We are met on a great battlefield of that war. We have come to dedicate a portion of that field as a final resting-place for those who here gave their lives that that nation might live. It is altogether fitting and proper that we should do this. But in a larger sense, we cannot dedicate, we cannot consecrate, we cannot hallow this ground. The brave men, living and dead who struggled here have consecrated it far above our poor power to add or detract. The world will little note nor long remember what we say here, but it can never forget what they did here. It is for us the living rather to be dedicated here to the unfinished work which they who fought here have thus far so nobly advanced. It is rather for us to be here dedicated to the great task remaining before us – that from these honoured dead we take increased devotion to that cause for which they gave the last full measure of devotion – that we here highly resolve that these dead shall not have died in vain, that this nation under God shall have a new birth of freedom, and that government of the people, by the people, for the people shall not perish from the earth.'

None of the journalists present at the dedication of the new national cemetery deemed Lincoln's words worthy of reporting. One wrote at the end of his account: 'The President also spoke.'

The Winds of War

1. *The Lost Army of Cambyses*: In 523 BC, the Persian emperor Cambyses sent 50,000 men across Egypt's western desert to destroy the oracle of Amun at Siwa. The army vanished, presumably swallowed by a great sandstorm.

2. *The Winds of Carthage*: In 262 BC the Roman invasion fleet of Marcus Atilius Regulus was destroyed by a storm on its way to Africa with the loss of 100,000 lives.

3. *The Divine Wind* (*Kamikaze*): Typhoons destroyed the Mongol invasion fleets that attacked Japan in 1274 and 1281. The first comprised 900 ships and 40,000 troops. The second was made up of 4,000 vessels and 140,000 troops.

4. *The English Winds*: The storm that scattered the Spanish Armada fleet in 1588 caused Philip II to lament that 'God is an Englishman'.

From 'A Soldier's Simple Recipes for Cooking in Trenches and Billets', by J. Noel (First World War)

BULLY BEEF STEW

Half a pound of bully beef, 2 or 3 potatoes (or any vegetables it is possible to procure. If possible, 1 onion).
Chop up potatoes and if you can get the onion fry it first in butter or bacon fat and add to the potatoes.
Cut up the beef into square pieces and place all in a mess tin.
Pour in a pint of water and some Symington's soup powder or Oxo cube and let boil gently for an hour.

BACON AND POTATOES

Peel and boil the potatoes first, then cut them into slices and fry slowly in cover of mess tin with the bacon.

BAKED POTATOES

Place the potatoes in the hot ashes of a fire. Cover over well and allow to bake.

TRENCH CAKE

Crush 4 or 5 army biscuits into powder. Add enough water to make a stiff paste. Mix in sugar and a tiny pinch of salt with a tablespoonful of butter if available.
Knead it well but not too heavily. Bake on a flat, hot stone which has been heated in a fire. (If you can use a beaten egg instead of the water it will make the cake much more tasty and light.)

JAM ROLL

4 or 5 army biscuits crushed into a powder and mixed into a stiff paste with water. Roll it out thin. Spread with jam and roll it up. Place it on a hot stone till brown.

TRENCH SAVOURY

Cut some bread thin and pour over some bacon dripping.

The Kamikaze's Pre-Battle Ritual

1. Don white headscarf with rising sun motif.
2. Wind around waist the belt of 1,000 stitches – crafted by a thousand women making one stitch each.
3. Drink a cup of sake.
4. Compose and recite death poem – traditional for samurai prior to hara-kiri.

Military Ranks

United Kingdom	USA
ARMY	
Field Marshal	General of the Army
General	General
Lieutenant General	Lieutenant General
Major General	Major General
Brigadier	Brigadier General
Colonel	Colonel
Lieutenant Colonel	Lieutenant Colonel
Major	Major
Captain	Captain
Lieutenant	First Lieutenant
Second Lieutenant	Second Lieutenant
Warrant Officer Class I	Chief Warrant Officer CW-5
Warrant Officer Class II	Chief Warrant Officer CW-4
Staff Sergeant ('Colour	Chief Warrant Officer CW-3
Sergeant' in Royal Marines)	Chief Warrant Officer CW-2
Sergeant	Warrant Officer WO-1
Corporal	Sergeant Major of the Army,
Lance Corporal	Command Sergeant Major,
Private ('Marine' in Royal	and Sergeant Major
Marines)	First Sergeant, Master
	Sergeant
	Sergeant First Class
	Staff Sergeant
	Sergeant
	Corporal, Specialist
	Private First Class
	Private

NAVY

Admiral of the Fleet	Fleet Admiral
Admiral	Admiral
Vice Admiral	Vice Admiral
Rear Admiral	Rear Admiral Upper Half
Commodore	Rear Admiral Lower Half
Captain	Captain
Commander	Commander
Lieutenant Commander	Lieutenant Commander
Lieutenant	Lieutenant
Sub Lieutenant	Lieutenant Junior Grade
Midshipman	Ensign
Warrant Officer	Midshipman
Chief Petty Officer	Chief Warrant Officer 4
Leading Rating	Chief Warrant Officer 3
Able Rating	Chief Warrant Officer 2
Ordinary Rating	Master Chief Petty Officer of the Navy, Master Chief Petty Officer
	Senior Chief Petty Officer
	Chief Petty Officer
	Petty Officer First Class
	Petty Officer Second Class
	Petty Officer Third Class
	Seaman, Airman, Fireman, Constructionman
	Seaman Apprentice, Airman Apprentice, Fireman Apprentice, Constructionman Apprentice
	Seaman Recruit, Airman Recruit, Fireman Recruit, Constructionman Recruit

AIR FORCE

Marshal of the Royal Air Force	General of the Air Force
Air Chief Marshal	General
Air Marshal	Lieutenant General
Air Vice Marshal	Major General
Air Commodore	Brigadier General
Group Captain	Colonel
Wing Commander	Lieutenant Colonel
Squadron Leader	Major
Flight Lieutenant	Captain
Flying Officer	First Lieutenant
Pilot Officer	Second Lieutenant
Acting Pilot Officer	Chief Master Sergeant of the
Warrant Officer	Air Force, Command Chief
Flight Sergeant, Chief	Master Sergeant, and Chief
Technician	
Sergeant	Master Sergeant
Corporal	Senior Master Sergeant
Senior Aircraftman	Master Sergeant
and Junior Technician	Technical Sergeant
Leading Aircraftman	Staff Sergeant
Aircraftman	Senior Airman
	Airman First Class
	Airman
	Airman Basic

Various Alliances

NATO – North Atlantic Treaty Organization (founded in 1949; membership as of 2003)

Belgium
Canada
Czech Republic
Denmark
France (withdrew from the
military structure in 1966)
Germany
Greece
Hungary
Iceland

Italy
Luxembourg
Netherlands
Norway
Poland
Portugal
Spain
Turkey
UK
USA

SEATO – South-East Asia Treaty Organization (1954–77)

Australia
France
Great Britain
New Zealand
Pakistan
The Philippines
Thailand
USA

The Warsaw Pact (1955–91)

Albania (left in 1961)
Bulgaria
Czechoslovakia
East Germany
Hungary
Poland
Romania
USSR

The Arab League (1945–)

Algeria	Morocco
Bahrain	Oman
Comoros	Palestine (PLO)
Djibouti	Qatar
Egypt*	Saudi Arabia*
Iraq*	Somalia
Jordan*	Sudan
Kuwait	Syria*
Lebanon*	Tunisia
Libya (withdrew 2002)	UAE
Mauritania	Yemen*

*founder members

ECOWAS – Economic Community of West African States (1975–)

Benin	Liberia
Burkina Faso	Mali
Cape Verde	Niger
Gambia	Nigeria
Ghana	Senegal
Guinea	Sierra Leone
Guinea-Bissau	Togolese Republic
Ivory Coast	

The Anzus Pact (1951–84)

Australia
New Zealand
USA

Words and Phrases of Warfare

'**Go the full nine yards**': When a Second World War fighter pilot loosed off all his ammunition – 9 yards of it – in one burst.

'**Dressed up to the nines**' After the British 99th Foot regiment, whose officers were said to be particularly well dressed in the eighteenth century.

'**Freeze the balls off a brass monkey**': Cannon balls were once kept in a brass rack (or monkey) which contracted when cold, ejecting the balls.

'**Pyrrhic victory**': Named after the ancient Greek king Pyrrhus of Epirus, who won a series of victories against the Romans that gradually depleted his veteran troops until they could no longer be replaced and he was finally defeated at Beneventum in 275 BC.

'**Hip, hip, hooray!**': Alleged to have been shouted by the Crusaders to mean 'Jerusalem is lost to the infidel, and we are on our way to paradise' – HIP or HEP being an acronym for *Hierosolyma est perdita*.

'**Gung ho**': From the Chinese for 'working together'.

'*Casus belli*': Cause of war.

'**Learning the ropes**': New sailors getting used to the rigging on sailing ships.

'**Scuttlebutt**': The cask of drinking water on board a ship, around which sailors would gather and gossip.

'**Show one's true colours**': From men-of-war that would approach an enemy ship while flying a friendly flag, then hoist their true flag immediately prior to opening fire.

'**Son of a gun**': In the days when women were allowed to live aboard naval vessels, children born at sea of unknown fathers were entered into the ship's log as 'son of a gun'.

'**Filibuster**': The name given to an individual waging war without government sanction in mid nineteenth-century Latin America.

'**Freelance**': A medieval knight without a lord or lands of his own.

'**Minutemen**': Colonial militia during the American War of Independence, so called for their instant readiness.

'**Turncoat**': A Duke of Saxony whose lands bordered France supposedly once dressed his men in blue coats that had a white interior, one to which they could switch when he wanted them to be thought to be acting in the French interest.

'**Under the yoke**': The Romans would force the troops of a defeated army to pass under a yoke (archway) of three spears to demonstrate their submission.

'**Vandal**': From the Teutonic tribe that sacked Rome for two weeks in 455.

'**Gazetted**': To have news of one's award for bravery published in the *London Gazette*.

'**V for Victory**': The two-fingered victory salute originates from the gesture used by English archers to taunt the French at Agincourt in 1415 – captured yeomen would have their index and middle fingers amputated so that they could never draw a bowstring again.

US Presidents of Military Rank (in order of presidency)

George Washington – Lieutenant General*
James Madison – Colonel
James Monroe – Lieutenant Colonel
Andrew Jackson – Major General
William H. Harrison – Major General
John Tyler – Captain (militia)
Zachary Taylor – Major General
Franklin Pierce – Brigadier General
James Buchanan – Private
Abraham Lincoln – Captain (militia)
Andrew Johnson – Brigadier General†
Ulysses S. Grant – General of the Army
Rutherford B. Hayes – Major General
James A. Garfield – Major General
Chester A. Arthur – Quartermaster General
Benjamin Harrison – Brigadier General
William McKinley – Major
Theodore Roosevelt – Colonel

Harry S. Truman – Major
Dwight D. Eisenhower – General of the Army
John F. Kennedy – Lieutenant (Navy)
Lyndon B. Johnson – Lieutenant Commander (Navy)
Richard M. Nixon – Lieutenant Commander (Navy)
Gerald R. Ford – Lieutenant Commander (Navy)
Jimmy Carter – Lieutenant (Navy)
Ronald W. Reagan – Captain
George Bush Snr. – Lieutenant (Navy)
George W. Bush – Lieutenant (National Guard)

* posthumously promoted six-star General of the Armies by Congress in 1976
† As military governor of Tennessee

Military Acronyms and Abbreviations

ABM – Anti-Ballistic Missile
AEF – American Expeditionary Force
AMRAAM – Advanced Medium Range Air-to-Air Missile
ANZAC – Australia and New Zealand Army Corps
ASDIC – Anti-Submarine Detection and Identification
    Commission
AWACS – Airborne Warning and Control System
AWOL – Absent Without Official Leave
BEF – British Expeditionary Force
CIA – Central Intelligence Agency
CIGS – Chief of the Imperial General Staff
CINCUS – Commander-in-Chief, US Fleet (abolished in 1941
    when its proximity to 'sink us' was noticed)
CTBT – Comprehensive Test Ban Treaty
GI – Has three derivations from the legends stamped on several
    pieces of US Army issue kit during the Second World War:
    Government Issue, General Issue, Galvanized Iron.
GCHQ – Government Communications Headquarters
DEFCON – Defence Condition
DMZ – Demilitarized zone

HALO – High Altitude Low Opening parachute jump

ICBM – Intercontinental Ballistic Missile

KGB – Committee for State Security (USSR) (Formerly NKVD)

KIA – Killed in Action

LAW – Light Anti-tank Weapon

MASH – Mobile Army Surgical Hospital

Medevac – Medical evacuation

MIA – Missing in Action

MOAB – Massive Ordnance Aerial Bomb (Mother Of All Bombs, colloquially)

MRE – Meals Ready to Eat

NAAFI – Navy, Army and Air Force Institute

NKVD – People's Commissariat of Internal Affairs (USSR) (Became the KGB in 1953)

NORAD – North American Air Defense Command

OSS – Office of Strategic Services

PIAT – Projectile Infantry Anti-Tank weapon

RADAR – Radio Detecting and Ranging

SA – *Sturm Abteilung* (Storm Detachment)

SALT – Strategic Arms Limitations Talks

SAS – Special Air Service

SBS – Special Boat Service

SHAEF – Supreme Headquarters Allied Expeditionary Force

SLBM – Submarine-Launched Ballistic Missile

SMERSH – A contraction of '*Smyert Shpionam*' ('Death to Spies'), Soviet counter-intelligence organization disbanded in 1958

SOE – Special Operations Executive

SONAR – Sound, Navigation and Ranging

SOS – Save Our Souls (the letters were originally meaningless – chosen because they were easy to remember in Morse code)

SS – *Schutzstaffel* (Protection Squad)

SSM – Surface-to-Surface Missile

START – Strategic Arms Reduction Talks

UNPROFOR – United Nations Protection Force

UNSCOM – United Nations Special Commission

Joseph Heller's Catch-22 (from the book of the same name)

'There was only one catch and that was Catch-22, which specified that a concern for one's safety in the face of dangers that were real and immediate was the process of a rational mind. Orr was crazy and could be grounded. All he had to do was ask, and as soon as he did, he would no longer be crazy and would have to fly more missions. Orr would be crazy to fly more missions and sane if he didn't, but if he was sane he had to fly them. If he flew them he was crazy and didn't have to; but if he didn't want to he was sane and had to.'

Military Mascots and Pets

1. BOY: A white dog that accompanied Prince Rupert into battle during the English Civil War and brought good luck to the Cavaliers. The Roundheads regarded him as an evil spirit and celebrated when he was killed in action at Marston Moor in 1644.

2. SATAN: A greyhound that saved a vital French position at Verdun in 1916 by bringing carrier pigeons to help direct defensive artillery fire despite having been shot by an enemy sniper.

3. PRINCE: An Irish terrier that went missing from his London home and tracked down his master, Private James Brown of the North Staffordshire Regiment, in Armentières, France, a few weeks later in 1914.

4. BLONDI: The Alsatian dog beloved of Adolf Hitler whom the Führer like to teach tricks.

5. ZUCHA: A Stakhanovite Russian dog that 'inspired other hounds' by finding 2,000 mines in eighteen days during the Second World War.

6. VOYTEK: A Syrian bear and mascot of the 2nd Polish Transport Company that trapped an enemy spy attempting to raid the company's ammunition supply in Iraq in 1942.

7. BUCEPHALUS: The steed of Alexander the Great. Alexander was the only man who could tame the horse, which he did after

noticing that it was afraid of its own shadow and turned its head towards the sun. When, at 30 years of age, Bucephalus died of wounds after carrying his master to safety, he was buried with full military honours.

8. INCITATUS: The Roman Emperor Caligula's horse, which the tyrant made a consul. Incitatus had an ivory manger and drank wine from a golden goblet.

9. THE HARTLEPOOL MONKEY: In 1805, the French ship the *Chasse Marée* was wrecked in a storm off the English coastal town of Hartlepool. When the vessel's mascot – a small monkey dressed in a miniature uniform – was washed up on shore, the local fisherman believed it to be a French spy and, and after holding an impromptu trial, summarily executed the creature by hanging.

'Cry havoc and let slip the dogs of war'

From *Julius Caesar*, by William Shakespeare, Act III, scene I

O, pardon me, thou bleeding piece of earth,
That I am meek and gentle with these butchers!
Thou art the ruins of the noblest man
That ever lived in the tide of times.
Woe to the hand that shed this costly blood!
Over thy wounds now do I prophesy, –
Which, like dumb mouths, do ope their ruby lips
To beg the voice and utterance of my tongue,
A curse shall light upon the limbs of men;
Domestic fury and fierce civil strife
Shall cumber all the parts of Italy;
Blood and destruction shall be so in use,
And dreadful objects so familiar,
That mothers shall but smile when they behold
Their infants quarter'd with the hands of war;
All pity choked with custom of fell deeds:
And Caesar's spirit, ranging for revenge,

With Ate' by his side come hot from hell,
Shall in these confines with a monarch's voice
Cry 'Havoc', and let slip the dogs of war,
That this foul deed shall smell above the earth
With carrion men, groaning for burial.

World Wars

Conflicts in which all the great powers of the time were involved

1. The Thirty Years' War, 1618–48
2. The War of the Spanish Succession, 1702–14
3. The Seven Years' War, 1756–63
4. The Revolutionary and Napoleonic wars, 1791–1815
5. The First World War, 1914–18
6. The Second World War, 1939–45

The US Marines Battle Hymn

From the halls of Montezuma
To the shores of Tripoli,
We fight our country's battles
In the air, on land, and sea.
First to fight for right and freedom,
And to keep our honor clean,
We are proud to claim the title
Of United States Marines.

Our flag's unfurl'd to every breeze
From dawn to setting sun;
We have fought in every clime and place
Where we could take a gun.

In the snow of far-off northern lands
And in sunny tropic scenes,
You will find us always on the job –
The United States Marines.

Here's health to you and to our Corps
Which we are proud to serve;
In many a strife we've fought for life
And never lost our nerve.
If the Army and the Navy
Ever gaze on Heaven's scenes,
They will find the streets are guarded
By United States Marines.

A Short History of Chemical Warfare

The use of chemical warfare in the First World War was pioneered by Fritz Haber, a Prussian Jew who went on to win the Nobel Prize for Chemistry in 1919 'for the synthesis of ammonia from its elements'.

On 22 April 1915 Haber was at the front lines directing the first gas attack in military history. About 150 tons of chlorine blew across the fields of Flanders, Belgium, spreading panic and death among the British and French soldiers. Haber returned to Berlin with his new technology apparently vindicated. A few days later, on 15 May, his wife Clara, appalled by his 'perversion of science', shot herself. Haber later invented Zyklon-B – the gas used by the Nazis to effect the Holocaust.

Some of the instances of chemical warfare

1. *Tear gas*: In August 1914, the first month of the First World War, the Germans claimed that the French fired tear gas grenades at their positions. In fact the French police (not army) had used tear gas before the war.

2. *'Sneezing powder'*: In the capture of Neuve Chapelle in October 1914 the German Army fired shells at the French which contained a chemical irritant that induced violent sneezing fits.

3. *Chlorine gas*: The first use of chlorine gas was on 22 April 1915, at the start of the Second Battle of Ypres. Used by the Germans against French and Algerian troops, the gas created a 4-mile gap in Allied lines.

4. *Phosgene*: First used in December 1915 by Germany. Its effects could be delayed for up to two days. Soldiers often inhaled lethal doses without realizing they had been gassed.

5. *Mustard gas*: First used by Germany against the Russians at Riga in September 1917. Unlike its predecessors, mustard gas did not have to be inhaled. An almost odourless chemical, it caused serious blisters both in the lungs and on any moist skin. Mussolini's victory in Abyssinia in 1937 was aided by the use of mustard gas bombs dropped from aircraft.

6. *Tabun*: The first nerve agent was discovered by German chemist Gerhard Schrader in 1936. He almost died after his assistant spilled a single drop of Tabun.

7. *VX nerve gas*: After its discovery by the British in 1952, the manufacture of VX began in quantity in the USA in 1961. Production ceased in 1968, when an accident at the plant in Dugway, Utah caused a cloud of the agent to be blown towards a nearby town, killing 6,000 sheep.

8. *Agent Orange*: Used along with Agent Purple, Agent Blue and Agent White by US forces during the Vietnam War 1965–75 to defoliate the vegetation surrounding the enemy.

9. *Chemical and nerve gas*: In 1988, Iraqi jet fighters dropped chemical and nerve gas on Halabja after the village fell into the hands of Kurdish rebels. Over 5,000 people died.

10. *Sarin*: In 1995, Japan's Aum Shinrikyo religious cult released sarin nerve gas in Tokyo's subway system during morning rush hour. Eleven people died and over 5,500 were injured.

One-Sided Victories

1. THE BATTLE OF MARATHON, 490 BC: When 10,000 Athenians defeated the 25,000 Persians of Darius the Great on the Plain of Marathon, they lost only 192 men to the enemy's 6,400.

2. THE TEARLESS BATTLE, 368 BC: Spartan King Archidamnus III routed a combined force of Argives and Arcadians without a single loss among his troops.

3. CANNAE, 216 BC: 80,000 Romans under Consuls Varro and Paulus were encircled by Hannibal's Carthaginian army. Up to 70,000 were killed in the republic's heaviest defeat. Hannibal lost only 5,700 men.

4. THE BATTLE OF MAGNESIA, 190 BC: Lucius Cornelius Scipio invaded Asia Minor with 30,000 troops to be met by a Seleucid force under Antiochus, numbering 72,000 (supported by scythed chariots and 54 elephants), of which 53,000 were killed for a loss of only 350 Romans.

5. ICENI REVOLT: According to Tacitus, in AD 61 a Roman army of 10,000 put down Queen Boudicca's rebellion, losing 400 men while slaughtering 80,000 Britons.

6. THE BATTLE OF ARGENTORATUM, 357: 13,000 Romans under Julian defeated 35,000 German Alamanni tribesmen, killing 6,000 Alamanni for the loss of only 247 Romans.

7. THE BATTLE OF JACINTO, 1836: 783 men led by General Sam Houston attacked and defeated 1,500 Mexicans who had been ordered to take a siesta by General Antonio Lopez de Santa Anna. Although the battle lasted only eighteen minutes, 630 men of the Mexican army were killed, for the loss of only nine Texans.

8. THE SHORTEST EVER WAR: In 1896 a usurper seized the throne of Zanzibar for precisely 45 minutes. A naval bombardment from three British warships destroyed the Sultan's palace and the usurper fled.

9. THE BATTLE OF OMDURMAN, 1898: An Anglo-Egyptian force of 26,000 men armed with machine guns and artillery met an army of 40,000 Mahdist Sudanese armed with swords and lances. The latter suffered 10,000 killed, 10,000 wounded and 5,000 taken prisoner, while the British had only 500 casualties.

10. THE PERSIAN GULF WAR, 1991: US and coalition forces lost 239 dead while inflicting over 100,000 casualties on the Iraqi Army during the liberation of Kuwait.

Ten of the Bloodiest Wars

1. THE SECOND WORLD WAR, 1939–45: *c.* 45m
   (including deaths in the Holocaust)

2. THE T'AI-P'ING REBELLION, 1851–64: *c.* 30 million deaths.
   The bloodiest ever civil war claimed between 20–40 million Chinese lives. Hung Hsiu-ch'uan, the defeated rebel leader, believed himself to be Jesus Christ's younger brother.

3. THE MANCHU-CHINESE WAR, 1644–90s: *c.* 25m.
   Marked the end of the Ming dynasty.

4. THE FIRST WORLD WAR 1914–18: *c.* 15m

5. THE NAPOLEONIC WARS, 1792–1815: *c.* 5m
   (including the French Revolution)

6. THE THIRTY YEARS' WAR, 1618–48: *c.* 4m

7. THE LOPEZ WAR, 1864–70: *c.* 2m.
   During the war between Paraguay and the Triple Alliance of Argentina, Brazil and Peru, Paraguay's population was reduced from 1,337,000 to 221,000.

8. THE SUDANESE CIVIL WAR, 1983–: *c.* 1.9m.
   One in five southern Sudanese have died as a result of the conflict and the associated famine.

9. THE CONGOLESE CIVIL WAR, 1998–: *c.* 1.7m.
   Between 1998 and 2003, an estimated 200,000 were killed in the fighting, with a further 1.5m succumbing to disease and starvation as a result of hostilities.

10. THE SEVEN YEARS' WAR, 1756–63: *c.* 1.4m.
    Between 1756 and 1962, the population of Prussia declined by half a million.

Mottoes

US Marine Corps	*Semper fidelis* (Always faithful)
US Army Corps of Engineers	*Essayons* (Let us try)
RAF	*Per ardua ad astra* (Through difficulties to the stars)
617 Squadron (Dambusters) RAF	*Aprés moi, le deluge* (After me, the flood)
Parachute Regiment (the 'Red Devils')	*Utrimque paratus* (Ready for anything)
Royal Army Medical Corps	*In arduis fidelis* (Faithful in adversity)
Special Air Service (SAS)	'Who dares wins'
US Army Special Forces	*De oppresso liber* (To free the oppressed)
All Scottish regiments (and the Scottish Crown)	*Nemo me impune lacessit* (Nobody insults me with impunity)
King's Regiment	*Nec aspera terrent* (Difficulties be damned)
Royal Irish Regiment	*Faugh-a-ballagh* (Clear the way)
King's Own Scottish Borderers	Once a Borderer, always a Borderer
Coldstream Guards	Second to None
US Air Force Security Force	*Defensor fortis* (Defender of the force)
Royal Marines	*Per Mare, Per Terram* (By Sea, By Land)
Royal Navy Submarine Service	We Come Unseen
Royal Gurkha Rifles	Better to Die than be a Coward
US Navy Admiral William F. Halsey, Jnr.	Hit hard, hit fast, hit often
West Point Military Academy	Duty, honour, country
Hitler's SS	*Meine Ehre heißt Treue* (My honour is loyalty)

Army Organization 2003

Corps	Two or more divisions
Division	From two to four brigades
Brigade	Three infantry battalions, one artillery regiment, one armoured regiment *or* two infantry battalions, two armoured regiments, one or two artillery regiments
Battalion	Three rifle companies, support company, head quarters company (Lt Col)
Company	Headquarters and three platoons (Major)
Platoon	Headquarters and three sections (Lt or 2Lt)
Section	Eight men (Cpl)

A regiment in the British infantry has no tactical significance (unlike the US, French and German armies where it roughly equates to a brigade); a British armoured or artillery regiment equates to a battalion.

'The Battle Hymn of the Republic' by Julia Ward Howe

Chorus:
Glory, Glory Hallelujah, Glory, Glory Hallelujah,
Glory, Glory Hallelujah, His truth is marching on.

Mine eyes have seen the glory of the coming of the Lord;
He is trampling out the vintage where grapes of wrath are stored;
He hath loosed the fateful lightning of His terrible swift sword,
His truth is marching on. (Chorus)

I have seen Him in the watchfires of a hundred circling camps;
They have builded Him an altar in the evening dews and damps;
I can read His righteous sentence by the dim and flaring lamps,
His day is marching on. (Chorus)

He has sounded forth the trumpet that shall never call retreat;
He is sifting out the hearts of men before His judgement seat,
Oh, be swift, my soul, to answer Him! Be jubilant, my feet!
Our God is marching on. (Chorus)

In the beauty of the lilies Christ was born across the sea,
With a glory in His bosom that transfigures you and me;
As He died to make men holy, let us die to make men free,
While God is marching on. (Chorus)

Countries with Mandatory National Service

(with no non-military alternative)

Afghanistan	Guinea-Bissau	North Korea
Albania	Honduras	Paraguay
Algeria	Georgia	Peru
Bolivia	Guinea	Philippines
Cambodia	Iran	Romania
Chile	Iraq	Singapore
China	Israel*	Somalia
Colombia	Kazakhstan	South Korea
Cuba	Laos	Sudan
Dominican	Lebanon	Thailand
Republic	Liberia	Tunisia
Ecuador	Libya	Turkey
Egypt	Madagascar	Venezuela
Equatorial Guinea	Mexico	Vietnam
Ethiopia	Mongolia	Yemen
Greece	Morocco	
Guatemala	Mozambique	

*In Israel, strictly orthodox Jews can find exemption from military service.

The Casualties and Costs of the First World War

Nation	Troops mobilized	Military dead	Military wounded	Civilian dead
France	8,410,000	1,357,000	4,266,000	40,000
British Empire	8,904,467	908,000	2,090,212	30,633
Russia	12,000,000	1,700,000	4,950,000	2,000,000
Italy	5,615,000	462,391	953,886	Unknown
USA	4,355,000	50,585	205,690	–
Belgium	267,000	13,715	44,686	30,000
Serbia	707,343	45,000	133,148	650,000
Montenegro	50,000	3,000	10,000	Unknown
Rumania	750,000	335,706	120,000	275,000
Greece	230,000	5,000	21,000	132,000
Portugal	100,000	7,222	13,751	–
Japan	800,000	300	907	–
Allied Total	42,188,810	4,887,919	12,809,280	3,157,633
Germany	11,000,000	1,808,546	4,247,143	760,000
Austria-Hungary	7,800,000	922,500	3,620,000	300,000
Turkey	2,850,000	325,000	400,000	2,150,000
Bulgaria	1,200,000	75,844	152,390	275,000
Central Powers Total	22,850,000	3,131,889	8,419,533	3,485,000
Grand Total	65,038,810	8,020,780	21,228,813	6,642,633

The Casualties and Costs of the Second World War

Nation	Troops mobilized (millions)	Military dead	Military wounded	Civilian dead
USA	14.9	292,100	571,822	–
UK	6.2	397,762	475,000	65,000
France	6	210,671	400,000	108,000
USSR	25	7,500,000	14,012,000	10-15,000,000
China	6–10	500,000	1,700,000	1,000,000
Germany	12.5	2,850,000	7,250,000	500,000
Italy	4.5	77,000	120,000	40-100,000
Japan	7.4	1,506,000	500,000	300,000
Others	20	1,500,000	–	14-17,000,000*
Total	105	15,000,000	25,028,822	26-34,000,000

* includes 6m European Jews and 4.5m Poles.

Impossible Odds

1. THE SPARTANS OF THERMOPYLAE, 480 BC: 300 warriors under Leonidas defended the pass of Thermopylae for two days against several hundred thousand soldiers from Xerxes I's Persian army. They inflicted a reported 20,000 casualties and gave their countrymen enough time to prepare their navies to fight and win the Battle of Salamis.

2. THE BATTLE OF JINGXING PASS, 204 BC: The fortified pass was held by the Zhao general Chen Yu with 200,000 soldiers. Han Xin routed them using only 12,000 of his troops, who succeeded in convincing the enemy that they were facing a much larger force.

3. THE BATTLE OF SHAYUAN, AD 537: The Chinese western Wei warlord Yuwen Tai led an army of 10,000 against the 200,000 men of his eastern rival Gao Huan, defeating them by killing 6,000.

4. THE BATTLE OF GUADALETE, 711: 12,300 Berbers and Arabs under Tarik ibn Ziyad defeated the Visigoth King Roderic's army of 90,000. The Jews of Toledo welcomed the Muslim conquerors as liberators.

5. THE NORMAN CONQUEST OF ENGLAND, 1066–70: William assembled an army of around 7–8,000 Normans, making a total of 10,000 including later arrivals, who went on to subdue a country of around 1.5 million people.

6. THE SIEGE OF DE-AN, 1206–7: The southern Chinese city was successfully defended by 6,000 troops under Wang Yunchu against 100,000 northern Jurchen attackers who employed siege towers and trebuchets.

7. THE MONGOLS: Between 1211–13, Genghis Khan conquered the Jin empire in China with 75,000 soldiers against 600,000.

8. THE BATTLE OF AUBEROCHE, 1345: During the Hundred Years' War, the Earl of Derby with 1,200 troops routed a French army of 7,000 that was besieging Auberoche in south western France, capturing the enemy commander Louis of Poitiers.

9. THE BATTLE OF KAUTHAL, 1367: 40,000 Muslim troops mustered by the Bahmani sultanate bested 540,000 Hindus from the kingdom of Vijayanagar by superior cavalry action when most of their fellows had already been routed.

10. THE BATTLE OF AGINCOURT, 1415: 900 men at arms and 5,000 archers commanded by Henry V of England defeated 20,000 French troops.

11. THE CONQUEST OF MEXICO: In 1521, Hernando Cortés landed with 550 Spaniards and went on to conquer an Aztec empire of 11 million subjects.

12. THE CONQUEST OF PERU: In 1530, Francisco Pizzaro landed with 150 Spaniards and captured the Inca capital Cajamarca, killing 7,000 enemy troops for no losses among his own men.

13. THE BATTLE OF ASSAYE, 1803: The Duke of Wellington, with 7,000 men and 20 cannon, prevailed against 75,000 Indian Marathas with 80 cannon.

14. **THE ALAMO, 1836:** Over a 13-day siege, 189 men – including Jim Bowie and Davy Crockett – fought to the death against over 2,000 Mexican troops under Santa Anna. In the process they inflicted over 1,000 casualties and bought time for the defence of Texas.

15. **RORKE'S DRIFT, 1879:** 150 British redcoats successfully defended a supply station against 4,000 Zulus. Eleven Victoria Crosses were awarded for the action.

16. **EAST AFRICA, 1914–18:** German General Paul von Lettow-Vorbeck's force of 3,000 European troops and 11,000 Africans remained undefeated by Allied armies totalling over 300,000 men over the course of the First World War. Though their rifle strength was only 1,400 at the time of the armistice, they inflicted 60,000 British and Indian casualties and an unrecorded number of African casualties.

Rome's Civil Wars

1. **THE ITALIAN SOCIAL WAR, 91–88 BC:** Rome's allies who were refused citizenship rebelled and were defeated.

2. **SULLA'S REVOLT, 88–82 BC:** Sulla turned his legions on Rome after a dispute with Marius, and won victory after the latter's death.

3. **CAESAR AND POMPEY, 49–45 BC:** When Julius Caesar declined to be prosecuted for his behaviour as Consul of Gaul and instead led his forces against the empire, Pompey the Great joined the battle against him. Pompey's armies fought on for three years after their master was murdered.

4. **ANTONY AND OCTAVIAN, 44–31 BC:** The armies of Mark Antony and Octavian defeated Brutus and Cassius – the co-conspirators in the murder of Julius Caesar. The two leaders then fell out, in part due to Antony's relationship with the Egyptian queen, Cleopatra. Octavian became the undisputed ruler of Rome after the Battle of Actium.

The Punic Wars, Rome vs Carthage

FIRST, 264–241 BC: After a series of sea battles, Rome gains control of Sicily.

SECOND, 218–201 BC: Hannibal defeats Rome at Trebia, Lake Trasimene and Cannae before losing to Scipio Africanus at Zama in 202 BC.

THIRD, 149–146 BC: After a three-year siege, the population of Carthage was sold into slavery and the city razed to the ground.

Nelson's Four Greatest Victories

Commemorated in bas-reliefs at the base of Nelson's Column in Trafalgar Square, London

1. CAPE ST VINCENT, 1797: Fought aboard HMS *Captain* against the Spanish off the southern coast of Portugal. A commodore at the time, Nelson was wounded in the stomach by flying timber and his vessel virtually wrecked in an action that saved the day for the British fleet. Yet he rallied his surviving marines and led a boarding party on to two adjacent enemy ships that had become entangled. The result was that fifteen British ships defeated a twenty-five-strong Spanish fleet.

2. THE NILE, 1798: Fought aboard HMS *Vanguard* against the French in the Bay of Aboukir on Egypt's Mediterranean coast. Nelson's fourteen vessels destroyed or captured eleven out of the thirteen enemy ships, though he was wounded in the forehead by shrapnel during the battle. He later wrote that 'Victory is not a name strong enough for such a scene'.

3. COPENHAGEN, 1801: Fought aboard HMS *Elephant* against the Danes. When ordered to break off his attack by Admiral Sir Hyde Parker, he famously claimed to be unable to see the signal. He pressed on with his division and broke the enemy's defensive line before persuading the Crown Prince of Denmark to agree to an armistice.

4. TRAFALGAR, 1805: Fought aboard HMS *Victory* against the French and Spanish at the mouth of the Mediterranean. Nelson split the enemy fleet in two, destroying one ship and capturing seventeen out of the thirty-three Allied vessels. Although Nelson died in the battle – shot down on his quarterdeck – Trafalgar was decisive and there were no further great fleet engagements in the Napoleonic Wars.

Arms and Armour of a Roman Legionary

Balteus	Military belt
Caligae	Hobnailed leather sandals
Cassis	Helmet
Focale	Scarf worn to protect the neck against chafing from armour
Gladius	Short sword
Lorica	Body armour for the torso
Manica	Segmented armguard
Ocreae	Shin-protecting greaves
Pilum	A heavy javelin 5–7 feet long
Pugio	Dagger
Scutum	Wooden shield with brass edging

The Wars of the Roses

Date	Battle	Victors
22 May 1455	First St Albans	Yorkists
23 Sept. 1459	Blore Heath	Yorkists
12 Oct. 1459	Ludford Bridge	Lancastrians
10 July 1460	Northampton	Yorkists
30 Dec. 1460	Wakefield	Lancastrians
2 Feb. 1461	Mortimer's Cross	Yorkists
17 Feb. 1461	Second St Albans	Lancastrians
28 Mar. 1461	Ferrybridge	Yorkists
29 Mar. 1461	Towton	Yorkists
25 Apr. 1464	Hedgeley Moor	Yorkists
15 May 1464	Hexham	Yorkists
26 Jul. 1464	Edgecote Moor	Lancastrians
12 Mar 1469	Losecote Field	Yorkists
14 Apr. 1470	Barnet	Yorkists
4 May 1471	Tewkesbury	Yorkist
22 Aug. 1485	Bosworth	Lancastrians (Tudors)
16 June 1487	Stoke	Lancastrians (Tudors)

Terrorist Organizations

Designated by the US Department of State

Abu Nidal Organization (ANO)
Abu Sayyaf Group
Al-Aqsa Martyrs Brigade
al-Jihad (Egyptian Islamic Jihad)
Al-Qa'ida
Armed Islamic Group (GIA)
Asbat al-Ansar

Aum Shinrikyo

Basque Fatherland and Liberty (ETA)

Communist Party of the Philippines/New People's Army
(CPP/NPA)

Gama'a al-Islamiyya (Islamic Group)

HAMAS (Islamic Resistance Movement)

Harakat ul-Mujahidin (HUM)

Hizbollah (Party of God)

Islamic Movement of Uzbekistan (IMU)

Jaish-e-Mohammed (JEM) (Army of Mohammed)

Jemaah Islamiya Organization (JI)

Kahane Chai (Kach)

Kurdistan Workers' Party (PKK) a.k.a. Kurdistan Freedom and
Democracy Congress (KADEK)

Lashkar-e Tayyiba (LT) (Army of the Righteous)

Lashkar i Jhangvi

Liberation Tigers of Tamil Eelam (LTTE)

Mujahedin-e Khalq Organization (MEK)

National Liberation Army (ELN)

Palestinian Islamic Jihad (PIJ)

Palestine Liberation Front (PLF)

Popular Front for the Liberation of Palestine (PFLP)

PFLP-General Command (PFLP-GC)

Real IRA

Revolutionary Armed Forces of Colombia (FARC)

Revolutionary Nuclei (formerly ELA)

Revolutionary Organization 17 November

Revolutionary People's Liberation Army/Front (DHKP/C)

Salafist Group for Call and Combat (GSPC)

Shining Path (Sendero Luminoso, SL)

United Self-Defence Forces of Colombia (AUC)

The Loads they Carried

Weight of personal weapons and equipment

Roman legionary under Marius, first century BC – 66lbs
Armoured French knight at Agincourt, 1415 – 80lbs
Union soldier at Gettysburg, 1863 – 50lbs
First World War American doughboy, 1917 – 48lbs
Allied infantryman on D-Day, 1944 – 80lbs
Russian soldier during the advance on Berlin, 1945 – 40lbs
Royal Marine yomping in the Falklands, 1982 – 120lbs
US Army soldier on patrol in Afghanistan, 2002 – 100lbs

The Payloads they Carried

Aircraft and their maximum bomb loads

AIRCRAFT	COUNTRY	PAYLOAD
Zeppelin VGO I Airship (1915)	Germany	2,200lbs
Boeing B-17 Flying Fortress (1935)	USA	4,000lbs
Avro Lancaster (1942)	Britain	14,000lbs
Boeing B-52 Stratofortress (1955)	USA	60,000lbs
Avro Vulcan (1956)	Britain	21,000lbs
Tu-160 Blackjack (1981)	USSR	36,000lbs
Rockwell B-1 Lancer (1985)	USA	75,000lbs
Northrop Grumman B-2 Spirit (1993)	USA	40,000lbs

Roman Legion Organization

 1 Contubernium = 8 men
10 Contubernia = 1 Century (80 men)
 2 Centuries = 1 Maniple (180 men)
 6 Centuries = 1 Cohort (480 men)
10 Cohorts and 120 horsemen = 1 Legion
   (5,240 men, including one extra-strength 'First Cohort')

Spears against Rifles – defeats of western armies by native forces

1. *Sudan, 1883* – The 'Mad Mahdi' and his dervishes massacred a force of 10,000 British troops under William Hicks.
2. *Isandlwhana, 1879* – 1,300 British and allied troops were wiped out by 20,000 Zulu warriors.
3. *Little Big Horn, 1876* – George Armstrong Custer and 264 men of the 7th Cavalry were slaughtered to the last man by Sioux Indian braves.
4. *Florida, 1816–42* – The Seminole Indians were the only tribe never to be defeated by the US Army. Holding out in the Everglades, they resisted forcible relocation to Mississipi and cost the government $20m and 1,500 lives in two wars (1816–23 and 1835–42).

Sticks against Swords – defeats of Roman Legions by ancient barbarians

1. *The Sack of Rome by Celtic Gauls*, 390 BC: The citizenry was alerted to the attack by the honking of the sacred geese of Juno, presaging a seven-month siege.
2. *The Battle of Allia*, 386 BC: The Roman army was routed by Gallic chieftain Brennus on the banks of the River Allia, north of Rome. Thereafter 18 July was regarded as unlucky.
3. *The Teutoburgian Forest*, AD 9: Three Roman legions under Quintilius Varus were slaughtered by Goths under Armininius, with survivors sacrificed to the barbarian gods and Varus's severed head sent to Roman leaders in the south.

4. *The Sack of Rome by Alaric's Visigoths*, AD 410.
5. *The Adrianople Campaign*, AD 376–378: In 378, the emperor Valens and 40,000 troops were slaughtered by the Goths at Adrianople.

Some Military Theorists

Sun Tzu (first half of fifth century BC): Author of *The Art of War*. Based on the premise that war is an evil, this is probably the best book on war ever written.

Aeneas the Tactician (mid third century BC): Greek author of the earliest work on military tactics. Of his oeuvre, only *On Siegecraft* survives.

Asclepiodotus (first century BC): Wrote *Outline of Tactics*, an essay on tactics containing a discussion of the structure of the Greek phalanx.

Onasander (mid first century AD): Writer of *The General*, a treatise on military psychology and tactics.

Sextus Julius Frontinus (late first century AD): His *Strategemata*, a compendium of military ruses, was famous throughout the Middle Ages.

Arrian (AD 86–160): Writer of textbooks on Greek and Roman military drill, including *Ektasis (Order of March)*.

Vegetius (late fourth century AD): His *Epitoma Rei Militaris (Epitome of Military Science)* covered all aspects of warfare from recruitment and training to naval combat.

Emperor Maurice (*c.* AD 600): Supposed author of *Strategikon*, a guide to all aspects of the Byzantine army of the seventh-century.

Jean de Bueil (*c.* 1410–70): Author of *Le Jouvencel*, a treatise on warfare as a political tool that predated Machiavelli's work.

Niccolò Machiavelli (1520–21): Author of *The Art of War* and, more famously, *The Prince*.

Chevalier Folard (1669–1752): Wrote *New Discoveries About War* and coined the phrase 'fog of war'.

Karl von Clausewitz (1780–1831): Prussian general and author of the justly famous *On War*, stressing the political psychology of warfare and the importance of logistics.

Mahan, Alfred Thayer (1840–1914): American naval historian and strategist who wrote *The Influence of Sea Power Upon History, 1660-1783*, advocating sea power as a determinant of the nation's strength.

J. F. C. Fuller (1876–1966): British pioneer of tank warfare.

Heinz Guderian (1888–1954): Writer of *Achtung Panzer!*, a development of the ideas of Fuller and Liddell Hart.

Basil Liddell Hart (1895–1970): Claimed to be the inventor of blitzkrieg tactics, which were ignored by the British authorities but taken up with enthusiasm in Germany.

Mao Tse Tung (1893–1976): Condensed the experience of the Long March into *On Guerilla Warfare*, 1937.

Military Orders

THE CHRISTIAN SOLDIERS OF THE MIDDLE AGES

1. *The Knights Templar*: Authorized by the papacy in 1128 to protect pilgrims on their way to bathe in the River Jordan. In 1307 their assets were seized by Philip IV of France, and their last Grand Master was burned for heresy in 1314.

2. *The Hospitallers of St John of Jerusalem*: Founded to look after sick pilgrims, they became a military order in the mid twelfth century. They ruled the island of Rhodes from 1306 to 1480, when they were expelled by the Turks. They retreated to Malta, which they fortified well enough to resist Suleiman the Magnificent's invasion of 1565, but were eventually defeated by Napoleon in 1798 after years of decay.

3. *The Teutonic Knights*: Amalgamated in 1212 from several Baltic orders, they ruled the region until their defeat by the Poles at Tannenberg in 1410.

4. *The Order of Our Lady of Ransom*: Founded in Aragon in 1218 by St Peter Nolasco for the redemption of captured crusaders.

5. *The Swordbearers*: Short-lived order founded by Albert, first Bishop of Riga, in 1197 for the conversion of Baltic pagans. Populated by adventurers rather than the pious, it was absorbed by the Teutonic Knights in 1238.

Nothern Irish Paramilitary Groups

IRA	Irish Republican Army (Republican)
PIRA	Provisional Irish Republican Army, formed after official IRA declared a ceasefire in 1972 (Republican)
RIRA	Real IRA, formed after PIRA declared a ceasefire in 1997 (Republican)
CIRA	Continuity IRA, attracted members from RIRA when that organization declared a ceasefire in 1998 (Republican)
INLA	Irish National Liberation Army (Republican)
SDA	Shankhill Defence Association (Loyalist)
UVF	Ulster Volunteer Force (Loyalist)
LVF	Loyalist Volunteer Force (Loyalist)
UDA	Ulster Defence Association (Loyalist)
UFF	Ulster Freedom Fighters (Loyalist) – Cover name for the UDA
RHD	Red Hand Defenders (Loyalist). Possibly does not exist, being only a cover under which UVF and UDA members can carry out acts of violence without breaking their ceasefire agreements.

'Sod 'Em All'

**Allied soldiers' ditty from the Second World War,
sung to the tune of 'Bless 'Em All'**

> Sod 'em all. Sod 'em all,
> The long and the short and the tall,
> Sod all the sergeants and WO1s,
> Sod all the corporals and their bastard sons,
> For we're saying goodbye to them all,
> As back to their billets they crawl,
> You'll get no promotion
> This side of the ocean,
> So cheer up, my lads, sod 'em all.

Notable Fortifications

1. **The Walls of Jericho:** Now Tel es-Sultan in Jordan, the Neolithic city possessed the earliest known stone walls and tower. Remains of the city date to the eighth millennium BC.

2. **The Great Wall of China:** Begun by the emperor Qin Shi Huangdi in 214 BC for defence against marauding nomad horsemen, the wall extends 4,000 miles westwards from Po Hai on the Yellow Sea. It is one of only two man-made constructions reputedly visible from outer space (the other being New York City's Staten Island garbage dump).

3. **The Long Walls of Athens:** Two parallel walls built to form a protected corridor 100m wide between Ancient Athens and the port of Piraeus that was essential for its supply in times of siege. Each was 6,500m long.

4. **Hadrian's Wall:** Built across Northern Britain by the Roman emperor Hadrian, the 1.8m/5.9ft thick wall ran 110km/68 miles from Bowness on the Solway Firth to Wallsend on the River Tyne. Fortified gateways ('milecastles') were located every Roman mile (1,480m/1,618yds).

5. **The Danevirke:** The 19km/12-mile rampart and ditch built across the base of the Jutland Peninsula around AD 800.

6. **The Ordos Loop Wall:** Built towards the end of the fifteenth century across the Ordos loop of the Yellow River to prevent Mongol attacks. 1,100km long, it included 800 forts and towers.

7. **The Hindenburg Line:** German First World War line of western fortifications constructed between 1916 and 1917, most of which were not breached until the British summer offensive of 1918.

8. **The Maginot Line:** Between 1929 and 1940, France built a line of concrete fortifications, tank traps and machine-gun posts supported by a network of bunkers and 100km of tunnels to guard against any future German invasion. In 1940, Hitler's troops bypassed these defences by going through Belgium, the Netherlands and the supposedly impassable Ardennes Forest.

9. **The Siegfried Line:** Built in the 1930s and strengthened later, this was Nazi Germany's last line of defence against the Western Allies. Also known as the West Wall, it ranged from heavily protected forts to 'dragon's teeth' tank obstacles and was not breached until the spring of 1945.

10. **The Berlin Wall:** Built in 1961 by the communist authorities to prevent East Germans from escaping to freedom in West Berlin.

11. **Hassan's Wall:** The late Moroccan King Hassan II built a wall of fortified rubble and razor wire over 1,000 miles long to protect his occupation of the Western Sahara after his 1975 invasion. Morocco spends $2m a day maintaining the wall and the occupying forces – paid for by the exploitation of Western Saharan oil wealth.

Napoleon's Marshals

Augereau	Jourdan	Murat
Bernadotte	Kellermann	Ney
Berthier	Lannes	Oudinot
Bessières	Lefebvre	Perignon
Brune	Macdonald	Poniatowski
Clauzel	Marmont	Serurier
Davout	Masséna	Soult
Gouvion St-Cyr	Moncey	Schet
Grouchy	Mortier	Victor

SAS Selection Procedure

Twice a year, up to 180 candidates volunteer from the regular army. The selection procedure involves:

1. Ten days of fitness and map-reading training in groups of twenty.
2. Ten days of solitary cross-country marching.
3. A 40-mile march in 20 hours carrying a 55-lb/25-kg rucksack.
4. At the end of the physical tests, candidates must be able to walk 4 miles in under 30 minutes and swim 2 miles in under 90 minutes.
5. Four weeks of weapons training.
6. Four weeks of jungle training.
7. Five-day escape and evasion exercise where candidates are hunted then interrogated.
8. Further courses in field medicine, signals, sniping, artillery spotting, demolitions, sabotage and languages.

Eight to ten candidates join the regiment.

Some Weapons and Equipment used by the SAS

1. Assault rifle: Eschewing the standard British Army SA80 rifle, most SAS troopers use Heckler and Koch MP-5 submachine guns or American M16A2s, some fitted with M203 grenade launchers.

2. Browning High Power 9mm pistol and Swiss-made Sig Sauer P226 semi automatic pistols.

3. GPMG ('Gimpy') or General Purpose Machine Gun, a precursor to the LSW (the standard issue heavy version of the SA80).

4. .50 Heavy Machine Gun made by Browning.

5. Mk19 40mm grenade launcher.

6. M72 LAW 66mm: A single shot anti-tank missile that penetrates up to 300mm of armour. Now being phased out.

7. 81mm L16 Mortar, which fires smoke, illuminating rounds or up to fifteen high explosive rounds per minute.

8. Milan wire guided anti-tank missile, which penetrates up to 1000mm of armour.

9. Stinger: A shoulder-fired anti-aircraft missile, with a maximum range of 4.5km.

10. Claymore mine: 1kg plastic explosive covered by 800 ball bearings, electrically detonated.

11. L9A1 Barmine: An anti-tank mine used for demolitions.

12. 'Flashbangs': Stun grenades.

13. Night vision goggles.

14. Thermal imaging viewers, for seeing inside buildings.

15. Land Rover 110, with a 3.5-litre V8 engine and two heavy-weapon mounts.

16. Global Positioning Satellite receivers for navigation.

17. Laser target designators for guiding air strikes.

18. Italian-made Agusta 109 helicopters (captured from the Argentinians during the Falklands War).

Top Fighter Aces

These figures are much disputed even today.
What follows are best estimates.

First World War

1. Manfred von Richthofen (the Red Baron) – Germany – 80 aircraft downed
2. René Paul Fonck – France – 75
3. William Bishop – Canada – 72
4. Raymond Collishaw – Canada – 62
5. Ernst Udet – Germany – 62

Top British ace – Edward Mannock – 61
Top US – Edward V. Rickenbacker – 26

Second World War

1. Erich Hartmann – Germany – 352
2. Gerhard Barkhorn – Germany – 301
3. Guenther Rall – Germany – 275
4. Otto Kittel – Germany – 267
5. Walter Nowotny – Germany – 258

Top Japanese ace – Hiroyoshi Nishizawa – 113
Top Russian – Ivan Kozhedub – 62
Top French – Pierre Clostermann – 33
Top Italian – Teresio Martinoli – 22
Top Polish – Stanislaw Skalski – 22
Top Norwegian – Sevin Heglund – 14
Top Dutch – Gerald Kesseler – 16
Top Hungarian – Dezso Szentgyorgyi – 34
Top Czech – Karel Miroslav Kuttelwascher – 20
Top Croat – Mato Dubovak – 40
Top Romanian – Constantine Cantacuzine – 60

Top female ace of the Second World War – and of all time – was Lydia Litvak of the USSR with 12 kills between 1941 and 1943

Top Royal Air Force Aces of the Second World War

1. Marmaduke 'Pat' Pattle – South Africa – 40+
2. James 'Johnnie' Johnson – UK – 36
3. Brendan 'Paddy' Finucane – Ireland – 32
4. George Beurling – Canada – 31
5. John Braham – UK – 29
6. Adolf 'Sailor' Malan – South Africa – 28
7. Clive Caldwell – Australia – 28
8. James 'Ginger' Lacey – UK – 28
9. Neville Duke – UK – 27
10. Colin Gray – New Zealand – 27

Top American Aces of the Second World War

1. Richard I. Bong – 40
2. Thomas McGuire – 38
3. David McCampbell – 34
4. Francis 'Gabby' Gabreski – 28
5. Robert S. Johnson – 27
   Charles MacDonald – 27
6. Joseph Foss – 26 *
   George Preddy – 26
7. Robert M. Hanson – 25
8. John C. Meyer – 24
   Cecil E. Harris – 24

* In January 2002, the 86-year-old Foss was detained and searched by airport security in Phoenix, Arizona after staff mistook his Congressional Medal of Honor for an offensive weapon and refused to believe that he had received it from President Roosevelt.

Korean War Aces

1. Nicolai V. Sutyagin – USSR – 21
2. Yevgeny G. Pepelyaev – USSR – 19
3. Lev Kirilovich Shchukin – USSR – 17
4. Joseph McConnell – USA – 16
5. James Jabara – USA – 15

'Ace of Aces' Erich Hartmann on Dogfighting

'Once committed to an attack, fly in at full speed. After scoring crippling or disabling hits, I would clear myself and then repeat the process. I never pursued the enemy once they had eluded me. Better to break off and set up again for a new assault. I always began my attacks from full strength, if possible, my ideal flying height being 22,000 feet because at that altitude I could best utilize the performance of my aircraft. Combat flying is based on the slashing attack and rough manoeuvring. In combat flying, fancy precision aerobatic work is really not of much use. Instead, it is the rough manoeuvre which succeeds.'

Notable Special Forces Missions

1. **Operation Claymore**, 1941: A British commando expedition to the Nazi-occupied Lofoten Islands off Norway destroyed eleven factories, 800,000 gallons of oil and five ships, acquired 314 volunteers for the free Norwegian forces and captured sixty collaborators, 225 German prisoners and a set of Enigma code ciphers. The only casualty was one self-inflicted thigh wound.
2. **Operation Oak**, 1943: In July 1943, Mussolini was removed from power by the king of Italy and imprisoned in a Gran Sasso ski hotel, which was accessible only by cable car. In September he was rescued in an airborne raid by German troops led by Otto Skorzeny.
3. **Operation Bricklayer**, April 1944: Major General Heinrich Kreipe, Commander of Germany's 22nd Infantry Division,

ESSENTIAL MILITARIA

was kidnapped from occupied Crete in a commando raid and spirited away by sea.

4. **Raid on Entebbe Airport**, 1976: A French airliner, hijacked by Palestinian and Baader-Meinhof terrorists, landed in Uganda with the collusion of despot Idi Amin. Israeli special forces flew 2,000 miles to the airport in four Hercules transports and rescued 103 hostages in 53 minutes. The Israelis lost only one soldier, shot in the back by a Ugandan sniper. Three hostages died in the assault, with a fourth later murdered by the Ugandans, while all seven hijackers, along with twenty Ugandan soldiers, were killed. The commandos also destroyed seven MiG jet fighters stationed at the airport.

5. **Mogadishu**, 1977: The German counter-terrorist unit GSG 9 and two British SAS men assaulted a hijacked Lufthansa airliner in Mogadishu in Somalia. Within six minutes, all 87 hostages had been rescued in the 'cleanest' special op ever.

6. **Moluccan Train Incident**, 1977: South Moluccan terrorists took 49 people hostage in a Dutch express train while a group of their comrades took 110 children hostage in an elementary school (releasing 106). The school hostages were rescued unharmed when Royal Dutch Marines burst through the wall of the building, catching three of their captors asleep. The train was assaulted by a second Marine team in a 20-minute attack while low-flying jet fighters ignited their afterburners overhead to distract the terrorists and keep the hostages' heads down. Two hostages were killed, with two marines and seven hostages wounded.

7. **Iranian Embassy Siege**, 1980: Six Iranians armed with sub-machine guns took control of the London embassy in protest at Ayatollah Khomeini's rule. Twenty-six hostages were rescued in 11 minutes by a team of twelve SAS troopers who killed five terrorists and captured the survivor.

8. **Operation Acid Gambit**, 1989: Robert Muse, an anti-Noriega agitator, was held hostage by the Panamanian regime in a milita-rized jail under armed guards who were instructed to kill him at the first sign of American aggression against the island. A Delta

Force team entered through the roof, killed or bound the guards (depending on whether or not they resisted), and rescued Muse with no losses, though two helicopters were shot down. Muse's bodyguard, despite having been hit in the head by a rotor blade, fended off enemy attacks until a US Army unit relieved the special forces.

9. **Sierra Leone**, 2000: SAS and British paratroopers rescued seven hostages from the West Side Boys armed militia in their jungle encampment. In an action that broke the West Side Boys as a force in the country, one of the rescuers was killed and one seriously wounded, while twenty-five militia men were killed and eighteen captured – including the leader.

Special Forces Disasters

1. **Cyprus Tragedy**, 1978: Egypt's TF777 special forces rescued thirty hostages from Arab terrorists claiming to represent the PLO in Nicosia, Cyprus. After completing the mission, fifteen of their number were shot dead by Cypriot National Guardsmen who mistook them for more terrorists, claiming that the Egyptian government had not forewarned them of the mission.

2. **Operation Eagle Claw**, 1980: After a group of Iranians took fifty-three men and women hostage in the US embassy in Teheran, a plan was devised under which a Delta Force team would land outside the capital in helicopters, make their way to the embassy, rescue the hostages and fight their way through a city of 6 million people to the airport, from where they would be flown out. The mission was aborted when one helicopter broke down en route and two collided in a dust storm, killing eight servicemen. One hostage later remarked, 'Thank God they never got here'.

3. **Malta Fiasco**, 1985: Palestinian terrorists hijacked an Egyptian Boeing 737, forcing it to land in Luqa, Malta. Lacking a plan of the internal layout of their own national carrier's aircraft, Egyptian special forces first blew a hole in the roof, killing twenty passengers in the rows beneath. They then burst into the cabin, firing wildly and hurling high-explosive grenades. Passengers who managed

to escape from the mêlée were mistaken for terrorists by TF777 troopers waiting outside and were shot down as they ran to safety. A total of fifty-seven hostages died.

4. **Afghanistan,** 2001: A US special forces soldier protecting the newly invested Afghan leader Hamid Karzai gave his own GPS co-ordinates to a B-52 bomber crew rather than those of a Taliban position. Karzai received only light wounds, but twenty-three Afghans and three Americans were killed.

5. **Moscow Theatre Siege,** 2002: When around eight hundred people were taken hostage by Chechen rebels in a Moscow theatre, Russian special forces pumped an unnamed gas into the building before their assault to incapacitate the terrorists and 116 hostages died from the effects of the chemicals.

Know Your Enemy

Admiral Yamamoto Isoroku's advice to junior officers of the Imperial Japanese Navy, 1939

'You can tell a man's character by the way he makes advances to a woman. Men like you, for example – when the fleet's in port and you go off to have a good time, you seem to have only two ways of going about things. First, you put it straight to the woman: "Hey, how about a lay?" Now, any woman, even the lowest whore, is going to put up at least a show of refusing if she's asked like that. So what do you do next? You either act insulted and get rough, or you give up immediately and go off and try the same thing on the next woman. That's all you're capable of. But take a look at Western men – they're quite different. Once they've set their sights on a woman, they invite her out for a drink, or to dinner, or to go dancing. In that way they gradually break down her defences until, in the end, they get what they want, and in style at that. Where achieving a particular aim is concerned, that's surely a far wiser way of going about things. At any rate, they're the kind of men you'd be dealing with if it were a war, so you'd better give it some thought.'

Famous Marches

1. THE MARCH OF THE TEN THOUSAND: A Greek mercenary force of veterans from the Peloponnesian War was hired by Cyrus the Younger to aid his rebellion against his elder brother in Persia. After Cyrus was killed at Cunaxa in 401 BC, the Greeks were forced to fight their way home in an action that became legendary for both their victories on the battlefields as they retreated and the atrocities they inflicted on the neighbouring populations.

2. HAROLD'S MARCH NORTH: In 1066, Harold Godwinson's Saxon army covered 190 miles to Tadcaster, near York, in less than five days. After defeating Harald Hardrada's Viking invasion force at Stamford Bridge on 25 September, Harold's army had to march back again, reaching London on 5 October, to face William and the Normans on the south coast. The exertion almost certainly contributed to their defeat at Hastings.

3. THE LONG MARCH: In 368 days from 1934–35, 90,000 Chinese Communists under Mao Tse-tung marched 6,000 miles/ 9,650km from Kiangsi to Yenan via Yünnan. Twenty-two thousand survived the journey and the continual attacks from Nationalist forces.

4. THE BIG YOMP: During the Falklands War of 1982, the British troops of 45 Commando marched 80 miles from San Carlos to Port Stanley in three days. Each man carried over 120lbs of equipment through hostile territory in freezing conditions.

The Effects of a One-Megaton Nuclear Airburst

By distance from detonation

12km	Many severely burned, all outdoors blinded, trees and buildings damaged. Winds reach 55kph/35mph.
8km	Most people severely burned, all outdoors blinded, trees blown down. Winds reach 150kph/95mph.
6.5km	Many dead from radiation, all outdoors blinded, most buildings damaged. Winds reach 260kph/160mph.
5km	Most people dead from burns, all outdoors blinded,

all houses destroyed, all larger buildings damaged. Winds reach 470kph/290mph.

3km   All people killed, all buildings destroyed. Winds reach 760kph/470mph.

The Haka

The Maori pre-battle chant now used by the New Zealand All Blacks rugby team before matches

Ka Mate! Ka Mate!	*It is death! It is death!*
Ka Ora! Ka Ora!	*It is life! It is life!*
Tenei te ta ngata puhuru huru	*This is the hairy person*
Nana nei i tiki mai	*Who caused the sun to shine*
Whakawhiti te ra	*Keep abreast! Keep abreast!*
A upane ka upane!	*The rank! Hold fast!*
A upane kaupane whiti te ra!	*Into the sun that shines!*
Hi!!	

Ten of the Bloodiest Battles

1. STALINGRAD, 1942–43: Around 300,000 German troops and 500,000 Soviet soldiers perished in Hitler's battle for the strategic industrial city on the River Volga.
2. THE KURSK SALIENT, 1943: In a series of engagements, including the largest tank battle in history, Germany suffered up to 70,000 casualties. The Soviet victors sustained up to one million casualties.
3. THE FIRST BATTLE OF THE SOMME, 1916: Between 1 July and 13 November, around 650,000 Germans, 420,000 British and 195,000 French became casualties. The British suffered 57,470 casualties on the first day alone, with one entire battalion – the 10th West Yorks – being virtually annihilated within a minute of their advance.
4. THE THIRD BATTLE OF YPRES, 31 July – 6 November 1917: 325,000 Allied and 260,000 German troops became casualties

at Passchendaele as Douglas Haig's offensive foundered under the heaviest rains in thirty years. The Germans regained most of their lost territory the following April.

5. VERDUN, 1916: 160,000 French troops died in the German assault on Verdun and the town's screen of forts along the River Meuse. German losses are thought to have been almost as high. By the end of the year, the French had recovered most of their ground.

6. THE BATTLE OF THE BULGE, 1944–45: Between 16 December and 25 January, Hitler's surprise counter-attack through the Ardennes inflicted 81,000 casualties on the Americans. However, the attack was blunted and the Germans suffered over 100,000 losses of their own.

7. OKINAWA, 1945: 12,000 Americans and 100,000 Japanese died in the struggle for the island.

8. BORODINO, 7 September 1812: In eleven hours of fighting described in Tolstoy's *War and Peace*, Napoleon lost 30,000 men and the Russians 45,000. The victory enabled Napoleon to enter Moscow unopposed, but enough Russians survived to bring about his eventual defeat.

9. GETTYSBURG, 1863: The bloodiest engagement on American soil resulted in 51,112 casualties (23,049 Union and 28,063 Confederate). The most casualties in one day during the US Civil War were suffered at Antietam on 17 September 1862, with 12,410 Union and 13,724 Confederate losses.

10. TOWTON, 1461: The most costly engagement on British soil took place during the Wars of the Roses. Total casualties in the Yorkist victory reached around 20,000.

Six Bloodless Battles

1. THE BATTLE OF BRÉMULE: On 20 August 1,119,900 English and French knights fought a pitched battle that resulted in only three deaths.

2. THE SIEGE OF SEVILLE: In 1078 King Alfonso VI of Castile prepared for a siege to drive out the Moors. When the Moorish

ruler, Al-Mutamid, heard that Alfonso was a chess enthusiast, he sent his chess champion, Ibn-Ammar, to play a game with the king for possession of the city. Ibn-Ammar won and the Castilians withdrew.

3. THE RECAPTURE OF CONSTANTINOPLE: In 1261 the Western Crusaders who had brought the city within the Latin empire in 1204 left it undefended while their fleet was at sea, allowing 500 Byzantine troops under Alexios to walk through the gates.

4. THE RECOVERY OF JERUSALEM: Frederick II led a crusade to the city in 1229, but won it though entreaty.

5. THE CAPTURE OF MOSCOW: On 14 September 1812 Napoleon took an undefended, emptied and burning city. The emperor himself was rescued from the blazing Kremlin by French looters and began the famous retreat on 19 October.

6. THE MOCK BATTLE OF MANILA: On 13 August 1898 General Jaudenes, the Spanish commander of the city's garrison, knew that his position was hopeless. However, to avoid the dishonour of surrendering to an American invasion force without a fight, he and Admiral Dewey choreographed an invasion whereby the Spanish would abandon successive sections of their defences according to signals flagged by the Americans.

Friendly Fire Incidents

1. Confederate General Stonewall Jackson was mistakenly and fatally shot by three of his own troops after the Confederate triumph at Chancellorsville in 1863.

2. In his first engagement during the First World War, Lawrence of Arabia shot his own camel in the back of the head.

3. Over the course of the First World War, up to 75,000 French troops were killed by their own artillery. The Germans suffered similar problems, their 49th Artillery Regiment being re-christened the $48\frac{1}{2}$ th for persistently firing short.

4. On 24 September 1915, in retaliation for German use of poison gas, 400 chlorine gas emplacements were established along the British front lines around Loos. After the gas was released, the

wind changed direction in the most northerly sector, blowing it back into the British trenches. Elsewhere, the gas was effective.

5. Italy's Marshal Italo Balbo, Mussolini's commander in Libya, was shot down by his own anti-aircraft defences at Tobruk in 1940.

6. The highest-ranking American to die in the Second World War was Lieutenant General Lesley McNair, killed by a stray bomb dropped by the US Army Air Corps.

7. Following a massive naval bombardment of the Aleutian island of Kiska in June 1943, 35,000 US and Canadian troops stormed ashore. Twenty-one troops were killed in the firefight before it was found there were no Japanese forces on the island.

8. A German ship used to intern Allied POWs lay anchored off Cape Bon for three days in May 1943. During that time, it was strafed and bombed by forty Allied aircraft. Poor shooting meant that only one POW was killed, with just a single bomb out of 100 – a dud – hitting its target.

9. In Vietnam in 1967, a US artillery unit selected the wrong powder charges for their ordnance, resulting in their shells falling on an American base camp – killing one and wounding thirty-seven. Believing they were under enemy attack, the gunners at the camp returned fire and during a 33-minute battle a further twelve were killed and forty wounded.

10. Fragging – In 1969, there were 126 incidents of American troops in Vietnam turning on their own officers, often by rolling a grenade into their tent. Thirty-seven men died as a result. In 1971 there were 333 incidents, resulting in twelve fatalities.

11. The 1991 Gulf War – A total of thirty-five of the 148 US combat deaths in the war were self-inflicted. A further seventy-eight US soldiers were wounded by their comrades, making friendly fire responsible for 17 per cent of all American casualties. Of the British troops involved in the conflict, nine were killed by US forces and only seven by the Iraqis.

Sacked Cities

1. THE CONQUEST OF KALINGA (now the Indian state of Orissa), 261 BC: After more than 100,000 of the region's inhabitants were killed by his invasion, the victorious but horrified Indian emperor Ashoka renounced violence and became a Buddhist.

2. THE FALL OF CARTHAGE, 146 BC: From a population of 200,000, only 50,000 survived to be enslaved by Scipio Aemilianus's Romans.

3. BERWICK, 1296: King Edward I of England, the Hammer of the Scots, spent three days slaughtering almost every adult, child and animal in the town. He eventually called off the massacre when he saw his men hacking a pregnant woman to death. 7,000–8,000 were killed.

4. KABUL, 1843: During their retreat from Kabul, 4,500 British and Indian soldiers and their families, together with 10,000 Afghans, attempted to flee the city and escape Dost Mohammed's forces. Only one man made it to the Indian border alive.

5. THE SACK OF NANKING, 1864: Government forces killed over 100,000 people in three days during the T'ai-p'ing Rebellion.

6. THE RAPE OF NANKING, 1937: 20,000 women between the ages of 10–70 were gang-raped by Japanese soldiers. Up to 200,000 men were executed.

7. LIDICE, 1942: Five days after the assassination of Deputy Gestapo Chief Reinhard Heydrich, the Czech village of Lidice was chosen at random for retaliation. The 172 men and boys over 16 were shot and the women and children separated and sent to concentration camps. Every building in the village was then torched and dynamited, with the remains bulldozed and then covered over with earth and concrete. Finally, new maps of the area were printed from which Lidice was excised.

8. ORADOUR-SUR-GLANE, 1944: When the SS Das Reich Division, rumoured to be carrying gold bullion to Normandy, was ambushed by the French Resistance, the nearby Oradour-sur-Glane became the arbitrary target for reprisals. On 10 June the division entered the town and turned the inhabitants out of their

homes. The men were taken away and shot, with many surviving to be burned alive. The women and children were locked in the local church, where a gas bomb was detonated. SS troops then entered with machine guns and grenades before finishing off the survivors by covering them with wood and kindling and setting it alight. Afterwards, the town was scoured for anyone who had been missed: an invalid who was burned in his bed and a baby, possibly hidden there, was roasted to death in a baker's oven. Six hundred and forty-two people died.

Assyrian Boasting

The destruction of Babylon: 'I levelled the city and its houses from the foundations to the top; I destroyed them and consumed them with fire. I tore down and removed the outer and inner walls, the temples and the ziggurats made of brick, and dumped the rubble in the Arahtu canal. And after I destroyed Babylon, smashed its gods and massacred its population, I tore up its soil and threw it into the Euphrates so that it was carried by the river down to the sea.' SENNACHERIB OF ASSYRIA, 680 BC

The firing of Nirbi: 'The city was exceedingly strong and was surrounded by three walls. The men trusted in their mighty walls and in their hosts, and did not come down, and did not embrace my feet. With battle and slaughter I stormed the city and captured it. Three thousand of their warriors I put to the sword; their spoil and their possessions, their cattle and their sheep I carried off. Many captives from among them I burned with fire, and many I took as living captives. From some I cut off their hands and their fingers, and from others I cut off their noses, their ears and their fingers, of many I put out the eyes. I made one pillar of the living, and another of heads, and I bound their heads to posts round about the city. Their young men and maidens I burned in the fire, the city I destroyed, I devastated, I burned it with fire and consumed it. At that time the cities of the land of Nirbi and their strong walls I destroyed, I devastated, I burned with fire.'

ASSUR-NÂSIR-PAL II, 883–859 BC

Frontline luxuries

1. **Prostitutes:** Napoleon introduced licensed brothels to tackle the gonorrhoea and syphilis rampant in the Grande Armée. The principle of forced medical inspections was enshrined in the Napoleonic Code of 1810, and authorized prostitutes serviced the French Army – and, occasionally, its allies – until the mid 1950s. It is rumoured that two prositutes were awarded medals for servicing an isolated garrison in the French Indo-China War of 1947–54.

2. **Absinthe:** Issued to French soldiers in Algeria from 1844 to 1847 as a malaria preventative.

3. **'Comfort women':** During the 1930s, around 200,000 Korean women were forced to work as sex slaves for the Imperial Japanese Army.

4. **Saunas:** Enjoyed by Finnish troops only 100m behind the front lines during the Russo–Finnish Winter War of 1939.

5. **Gourmet coffee:** Each Italian trooper in North Africa during the Second World War carried his own personal espresso maker.

6. **Twenty cigarettes a day:** Supplied to US troops in the Pacific Theatre during the Second World War.

7. **Fast food:** Burger King and Pizza Hut opened franchises for US troops waiting to invade of Iraq in 2003.

Drugs given to soldiers

1. *Vikings on mushrooms*: The Norse Berzerker warriors of the Middle Ages are thought to have eaten hallucinogenic fungi to induce battle rage.

2. *Germans on coke*: In 1883 Theodor Aschenbrandt administered cocaine to members of the Bavarian army. It was found that the drug enhanced their endurance on manoeuvre.

3. *Tommies on spirits*: British troops were often given an extra ration of rum before an advance during the First World War.

4. *Paratroopers on benzedrine*: Given to German airborne troops dropped into Crete in 1941.

5. *GIs on acid*: In 1957, 1,000 American servicemen, who volunteered to test gas masks at the Army Chemical Warfare Laboratories, were instead given doses of LSD to measure the drug's effects.

6. *Pilots on speed*: The US Air Force issues what it calls 'Go-pills' to personnel on extended missions. According to US Air Force investigators these amphetamines had been taken by the two pilots who bombed a squad of Canadian infantrymen on a night-time exercise in Afghanistan in 2002, killing four and injuring eight.

One-Man Armies

1. HORATIUS: The Roman sentry on the Tiber Bridge who in 508 BC held off the Etruscan army singlehandedly, long enough for his comrades to destroy the crossing.

2. CHORSAMANTIS THE AVAR: During the siege of Rome in AD 538, the warrior became maddened by drink and wounds and rode out alone to the barbarian camp. He was confronted by twenty enemy horsemen, whom he dispatched before being overwhelmed.

3. THE LONE VIKING: In 1066, Harold Godwinson's Saxon army marched to York to fight off Harald Hardrada's Norwegian invasion. Harold caught his enemies by surprise, but had to cross Stamford Bridge to get to them. However, the bridge was held by a single Viking champion who slew the first forty men who tried to advance. By the time a boat had been floated under the bridge and a long spear thrust upwards through the planks to kill him, the warrior had given his comrades enough time to ready their arms and armour and prepare their battle formations.

4. BENKEI (died 1189): The Japanese warrior monk stood on the Gojo Bridge in Kyoto and challenged all comers. According to legend he defeated 999 warriors in single combat before being beaten.

5. SIR WILLIAM MARSHAL (1146–1219): By common consent the greatest warrior of his age. His first engagement was the Battle

of Drincourt in 1167 where, though his warhorse was killed beneath him, he managed to defeat an estimated forty other knights in succession without pause.

6. PEDRO FRANCISCO (died 1831): The 6ft 6ins, 28olb Portuguese American was the most famous private soldier of the Revolutionary War. In 1779 Francisco captured the British flag at Stony Point, the British Army's stronghold on the Hudson River, and during one short engagement killed eleven enemy troops using his 6-foot-long broadsword. George Washington said that 'Without him we would have lost two crucial battles, perhaps the War, and with it our freedom. He was truly a One-Man Army'.

7. JUNIOR J. SPURRIER: The 22-year-old US Infantry Staff Sergeant was awarded the Congressional Medal of Honor in March 1945 for taking the town of Achain in northern France in a one-man assault. Using grenades, bazookas, heavy machine guns and his M1 rifle, Spurrier went from house to house, taking strongpoints one by one until he captured the garrison commander and nineteen others, killing twenty-five German soldiers in the process.

8. AUDIE MURPHY (1924–71): Murphy was turned away from the Marines and the Paratroopers for being only 5ft 5ins tall, yet went on to become the most decorated US soldier of the Second World War. The engagement for which he won the Medal of Honor took place near Holzwihr in France in January 1945. When his unit was attacked by six German tanks with waves of infantry in support, Murphy ordered his men to retreat into nearby woods while he remained behind to direct artillery fire. When his position was overrun, he climbed on to a burning tank destroyer and held off the enemy with .50-calibre fire. After an hour he had killed or wounded fifty German troops and forced the tanks to retreat. Murphy later enjoyed a successful acting career in Hollywood – playing himself in the 1955 movie *To Hell and Back* – before dying in a plane crash in 1971.

9. LACHHIMAN GURUNG: In May 1945 in Taungdaw, Burma, the Gurkha rifleman was manning the foremost foxhole of his unit's position when 200 Japanese troops attacked. The first two grenades to be thrown into his trench he picked up and threw back, but a third exploded in his right hand – blowing off his fingers, shattering his arm and permanently blinding him in one eye. For the next four hours, Gurung loaded and fired his rifle with his left arm, killing thirty-one enemy soldiers before the Japanese finally retreated. He remains one of the most celebrated winners of the Victoria Cross.

10. HIROO ONODA: Japanese soldier Lieutenant Onoda refused to stop fighting long after the Second World War was over, claiming that stories of the war's ending were mere propaganda. It wasn't until March 1974, when his former commanding officer flew out to the remote Pacific island where Onoda was dug in and ordered him to lay down his arms that he finally complied. However, his record was broken by Private Teruo Nakamura, who maintained his resistance on the island of Morotai for a further nine months until December of 1974.

Decorations and Awards won by Audie Murphy, First Lieutenant, 3rd Infantry Division

The most decorated US soldier of the Second World War

Medal of Honor
Distinguished Service Cross
Silver Star with First Oak Leaf Cluster
Legion of Merit
Bronze Star Medal with 'V' Device and First Oak Leaf Cluster
Purple Heart with Second Oak Leaf Cluster
US Army Outstanding Civilian Service Medal
Good Conduct Medal
Distinguished Unit Emblem with First Oak Leaf Cluster
American Campaign Medal

European–African–Middle Eastern Campaign Medal with One
  Silver Star
Four Bronze Service Stars (representing nine campaigns)
Bronze Arrowhead (representing assault landing in Sicily and
  Southern France)
Second World War Victory Medal
Army of Occupation Medal with Germany Clasp
Armed Forces Reserve Medal
Combat Infantry Badge
Marksman Badge with Rifle Bar
Expert Badge with Bayonet Bar
French Fourragère in colours of the Croix de Guerre
French Legion of Honour, Grade of Chevalier
French Croix de Guerre with Silver Star
French Croix de Guerre with Palm
Medal of Liberated France
Belgian Croix de Guerre 1940 Palm

Great Fleet Battles

1. SALAMIS, 480 BC: Between 310 and 368 Greek ships commanded by Eurybiades of Sparta defeated some 600 Persian
   vessels by trapping them in the Salamis Strait, where many were
   either outmanoeuvred and destroyed or driven ashore.
2. THE ARGINUSAE ISLES, 406 BC: In the greatest sea battle of
   the Peloponnesian War, 150 Athenian triremes defeated 120
   Spartan vessels, sinking 77 for the loss of 25 of their own.
   However, six of the eight Athenian generals involved in the
   engagement were later executed for failing to pick up survivors.
3. AEGOSPOTAMI, 405 BC: The final battle of the Peloponnesian
   War, in which the Athenian fleet was destroyed by the Spartans
   under Lysander.
4. ECNOMUS, 256 BC: 330 Roman quinqueremes fought 350
   Carthaginian warships. The day went to the Roman fleet, with
   losses of 24 and 94 respectively.
5. ACTIUM, 31 BC: Octavian's 400 ships defeated the combined

fleets of Mark Antony and Cleopatra with 500. Octavian became Augustus, the undisputed ruler of the Roman world.

6. LEPANTO, 1571: 33,000 died in the last great confrontation at sea to be fought by galleys. Each side had around 200 ships, the Holy League scoring a decisive victory over the Turkish fleet.

7. THE SPANISH ARMADA, 1588: Of the 130 ships that sailed from Lisbon to conquer England, only half returned in failure due to the attacks of the English fleet in the Channel and storms around the coasts of Britain.

8. THE FOUR DAYS' BATTLE, 1–4 June 1666: England's Admiral George Monck with 56 ships engaged the Dutch Admiral Michiel de Reuyter's 85 vessels off the North Foreland, Kent. The battle only ended when both sides had run out of ammunition, and although Monck received reinforcements of a further 24 ships on the 3rd, the English suffered a serious defeat, losing 17 ships and 8,000 casualties against the seven ships and 2,000 casualties suffered by the Dutch.

9. TRAFALGAR, 1805: Admiral Horatio Nelson famously split his twenty-seven warships and attacked the Admiral Pierre-Charles Villeneuve's combined French and Spanish fleet of thirty-three in two squadrons. Though he lost his life to a sniper's bullet, Nelson was rewarded by the capture of 20 French vessels and two sunk without any losses to his own force. One of the most decisive battles in history, it put an end to Napoleon's plans to invade Britain.

10. JUTLAND, 1916: The largest naval engagement of the First World War, in which 151 British ships clashed with ninety-nine German in the North Sea. The result was a strategic victory for the British, although they lost fourteen ships to the enemy's eleven and suffered over twice their number of casualties, prompting Admiral Sir David Beatty's remark, 'There seems to be something wrong with our bloody ships today'.

11. LEYTE GULF, 1944: Over five days and a vast area of sea, 218 American warships supported by 1,280 aircraft engaged 64 Japanese vessels with 716 warplanes in support. Twenty-six Japanese and six American ships were sunk.

Failed Technologies

1. THE BALLONKANONE: The first anti-aircraft weapon, a 37mm cannon designed by the Germans to shoot down French balloons during the Siege of Paris in 1870–71. None were shot down.

2. PATRIOT MISSILE: In the First Gulf War, forty-seven Patriots were fired, hitting only a single incoming Scud.

3. MAUS: A giant 192-tonne German tank built in 1945 that was too heavy for bridges and soft ground. Despite a 1,200-horsepower engine, its top speed was only 12mph.

4. FLYING SAUCER: A Canadian firm designed the VZ-9AV Avrocar 'flying saucer' for the US Air Force, but the shape proved unstable in flight.

5. MESSERSCHMITT ME 163 KOMET: This German fighter plane was powered by a rocket engine. It was used against Allied bombers from 1944 onwards, but its volatile fuel often tended to vaporize the pilot before he could engage enemy aircraft.

6. ROCKWELL XB-70 VALKYRIE: America's Mach-3 bomber was abandoned in 1961 as even at speeds of over 1,864mph/3,000 kmph it was vulnerable to surface-to-air missiles.

7. THE MITRAILLEUSE: A machine gun thought to be a war-winning weapon by the French was kept so secret that instructions for its use were only distributed on the opening day of the 1870 Franco-Prussian War, by which time they were too late.

8. THE NO. 74 HAND GRENADE: This British explosive was designed to stick to tanks, but was discontinued after it kept sticking to the hands of its throwers.

9. THE LUNGE BOMB: A Japanese anti-tank weapon consisting of a grenade on the end of a long spear. The act of placing the weapon in a tank's tracks was usually enough to detonate its charge before the user could retreat to a safe distance.

Rates of Fire down the ages

Arquebus / Matchlock musket (fifteenth century)	Two rounds per minute (rpm)
Wheel-lock musket (sixteenth century)	Two to three rpm
Flintlock musket (seventeenth century)	Three rpm
Gatling gun (1860s)	200 rpm
Maxim gun (1880s)	600 rpm
Lee-Enfield rifle (1900s)	Eight rpm
Thompson submachine gun (1920s)	725 rpm
Bren gun (1930s)	500 rpm
M1 Garand rifle (1940s)	24 rpm
Kalashnikov AK-47 (1940s)	600 rpm
M16 assault rifle (1950s)	800 rpm
GPMG (1950s)	1,000 rpm
M134 Minigun (1960s)	6,000 rpm
SA-80 assault rifle (1980s)	770 rpm
'Metal Storm' (1990s)*	1,000,000 rpm

* Fired by electronic ignition, with no firing mechanism.

From 'The Modern Traveller', by Hilaire Belloc

Blood thought he knew the native mind;
He said you must be firm, but kind.
A mutiny resulted.
I never shall forget the way
That Blood stood upon this awful day
Preserved us all from death.
He stood upon a little mound
Cast his lethargic eyes around
And said beneath his breath:
'Whatever happens we have got
The Maxim Gun, and they have not.'

Big Guns

1. **THE GREAT CANNON OF MEHMED:** A 42-inch bombard was used by the Turks to attack the walls of Constantinople with 1,200lb/543kg stone balls. Its range was 1 mile but it could be fired only seven times each day.

2. **THE TSAR PUCHKA:** Now on display in the Kremlin, the 40-ton 'King of Cannons' was built in the sixteenth century, with a bore of 36.2 inches and a barrel 10-feet long.

3. **BIG BERTHA:** A 16.53-inch mortar named after the manufacturer Gustav Krupp's wife, this 43-ton gun could fire a 2,200lb shell over 9 miles. Transported by Daimler-Benz tractors, it took its 200-man crew over six hours to re-assemble it before it was used to destroy the Belgian defences at Liège during the First World War.

4. **THE PARIS-GESCHÜTZ:** 'Paris Gun': Built by the Germans to shell the French capital during the First World War, it had a range of 80 miles/130km.

5. **THE 'BOCHEBUSTER':** An 18-inch train-mounted howitzer that could fire a 2,500lb/1,133kg shell up to 22,800 yds/20,850m was used by the British from 1940 to help defend the Kent coast.

6. **GUSTAV:** The 31.5-inch gun used by the Germans in the siege of Sebastopol in 1942 could fire a 10,500lb shell 29 miles/47km. It fired 300 rounds, including 48 in the Crimea, before its barrel was worn out.

7. **V3:** Built by the Germans in static underground firing tubes for the shelling of London. Never used, is range was 95 miles/153km.

8. **THE IRAQI SUPERGUN:** In 1988 construction began of a gun with a 1,000mm bore that could fire a 600kg projectile over 620 miles/1,000km. Work on the 'Babylon Gun' was halted by UN weapons inspectors after the Persian Gulf War of 1991.

Winners of the Dickin Medal for Animal Gallantry

'The Animals' VC' instituted by Maria Dickin in 1943

1. *Rob the 'Paradog'*: A mongrel that served with the SAS in North Africa in the Second World War and made over twenty parachute jumps.

2. *Ricky*: A Welsh sheepdog who continued to sniff out landmines despite having one explode in his face in Holland in 1944.

3. *GI Joe*: An American carrier pigeon that flew 20 miles/32km in 20 minutes just in time to stop a bombing raid on a village that had just been taken – thereby saving up to 100 Allied lives.

4. *Antis*: A German Alsatian acquired as a puppy from a bombed-out farmhouse by Czech airman Jan Bozdech when he was shot down behind enemy lines during the Second World War. Antis flew with Bozdech thereafter, being wounded twice. When Bozdech fell foul of post-war Czechoslovakia's communist authorities and had to flee the country, the dog saved his master's life by savaging a border guard.

5. *Mary of Exeter*: A carrier pigeon that flew from 1940 to the end of the Second World War, surviving three pellets in her body, a wing shot away, two bombing raids on her loft and an attack by a hawk requiring twenty-two stitches.

6. *Beauty*: A terrier that helped dig out sixty-three men and women and one cat from under rubble during the Blitz.

7. *Simon*: The ship's cat that served on board the British escort sloop HMS *Amethyst* in April 1949 in China as Mao Tse-Tung's forces took control of the country. When the *Amethyst* was trapped on the Yangtze River and shelled by the Chinese, Simon was trapped in the wreckage for four days. Despite his wounds, he continued to hunt rats and protect the crew's food supply throughout a summer-long siege. The story was made into a film, *The Yangtze Incident*, released in 1956.

Nations with the Smallest Armies

1. Costa Rica   none
2. Iceland   none
3. Antigua and Barbuda   150 troops
4. Seychelles   450
5. Barbados   610
6. Luxembourg   768
7. Gambia   800
8. Bahamas   860
9. Belize   1,050
10. Cape Verde   1,100

Top French Collaborators

1. MARSHAL HENRI PHILIPPE PÉTAIN: Great War hero turned Vichy leader.
2. RUDOLPHE PEUGEOT: Made tanks for Germany until forced by the French Resistance to destroy his factory in order to prevent further Allied bombing of the area.
3. THE MICHELIN FAMILY: Made high-quality tyres for the Wehrmacht. The RAF destroyed their factory when they refused to do so themselves.
4. COCO CHANEL: The designer was unrepentant about her relations with Nazi officers, declaring, 'My heart is French, but my cunt is international'.
5. ROBERT BRASILLACH: Intellectual turned Nazi.

Dogs of war – Mercenaries Past and Present

1. *Xenophon* (*c.* 430–350 BC): Greek writer who fought for Cyrus the Younger of Persia in his victorious rebellion against his brother Artaxerxes at the Battle of Cunaxa in 401 BC. After Xenophon's employer turned on his allies, he led his fellow Greeks home in a march recounted in his *Anabasis*.
2. *The Three Captains*: After Carthage's defeat in the First Punic War, she was unable to pay the arrears of her mercenary armies

led by Autaritus, Mathos and Spendius. In the Mercenary War of 241–237 BC, Tunis was seized, a revolt in Libya incited and Carthage besieged before the victory of Hamilcar Barca at Bagradas in 240 BC.

3. *Mercadier*: The brutal mercenary ally of Richard the Lionheart became notorious in France for maiming and massacring prisoners of war in the twelfth century.

4. *The Condottiere*: Mercenary companies that formed in the fourteenth century in the wake of the Hundred Years' War and operated in Italy. They included the Catalan Company composed of Catalans, the Grand Company composed of Germans and Hungarians and the White Company led by the Englishman Sir John Hawkwood.

5. *Georg von Frundsberg* (1473–1528): Called 'Father of the Lands-knechts', the Holy Roman Emperor Maximilian I's famous mercenary infantry. Frundsberg won most of Lombardy for Charles V at Bicocca in 1522 and helped defeat the French at Pavia in 1525.

6. *Giovanni Giustiniani*: The mercenary commander who marshalled 7,000 defenders against the 100,000 Turks who besieged Constantinople in 1453. Most of the citizens refused to aid his troops and later suffered for their timidity.

7. *The Flying Tigers*: American pilots recruited to fight the Japanese in Burma and China in 1941–42. In ten weeks of action over Rangoon, they inflicted 200 losses on superior Japanese forces while losing only sixteen of their own aircraft.

8. *Air America*: From 1955 to 1974, CIA-employed pilots who flew medevac and reconnaissance missions against the communist insurgents in Laos and transported supplies and troops.

9. *Bob Denard*: Denard left the French Navy when he was 20 after burning down a restaurant during a drinking bout. He embarked on a career that took him from Vietnam in the 1940s to the Belgian Congo in the 1960s, when he fought UN troops backing the Zairean leader, Patrice Lumumba. Once asked how he could fight for so many different causes, he explained, 'I'm not into politics'.

10. *'Mad' Mike Hoare*: A hard-drinking Irish-born colonel whose exploits in the Congo and South Africa in the 1970s inspired movies such as *The Wild Geese*. In 1982 he was found guilty of hijacking a plane to escape from a failed coup in the Seychelles and sentenced to ten years' imprisonment by a South African court. He and his men had infiltrated the islands disguised as a beer-tasting team, the Ancient Order of Frothblowers.

11. *Tim Spicer*: London-based Executive Outcomes was hired in 1995 to put down the Revolutionary United Front and keep the peace during the 1996–97 elections in Sierra Leone. A force of 200 South Africans performed the job successfully, but a violent coup followed swiftly when the IMF forced the mercenaries to leave. In 1998, ex-SAS officer Spicer's Sandline Security helped restore the elected President Kabbah while British Foreign Secretary Robin Cook denied his ministry's involvement.

The Normandy Beaches

Used in the D-Day Landings

Omaha – US landings
Utah – US
Juno – Canadian
Sword – British
Gold – British

Codenames for Military Actions

Sea Lion: Nazi invasion of Britain initially scheduled for September 1940.
Barbarossa: Hitler's invasion of Russia, 1941.
Fortitude: Allied plan to deceive Germany that the D-Day landings of 1944 would take place in Pas de Calais rather than Normandy.
Chastise: Dambusters raid, May 1943.
Husky: Allied invasion of Sicily, 1943.
Overlord: Allied invasion of continental Europe, 1944.

Market Garden: Allied airborne and ground attacks through Holland, September 1944.

Zapata: The landings on Cuba's Bay of Pigs, 1961.

Rolling Thunder: US bombing of North Vietnam, 1965–68.

Thunderbolt: Israeli commando raid on Entebbe Airport, 1976 (later renamed Operation Jonathan in memory of Jonathan Netanyahu, an officer who died on the mission).

Eagle Claw: US attempt to rescue hostages from Embassy in Teheran, 1980.

Corporate: British invasion of The Falklands, 1982.

Urgent Fury: US amphibious landings in Grenada, 1983.

Just Cause: US invasion of Panama, 1989.

Desert Storm: US and allied action to expel Iraqi forces from Kuwait, 1991.

Restore Hope: US peacekeeping in Somalia, 1992.

Uphold Democracy: US peacekeeping in Haiti, 1994.

Desert Fox: US and British bombing of Iraq, 1998.

Allied Force: Nato bombing of Serbia and Kosovo, 1999.

Enduring Freedom: Allied action against the Taliban in Afghanistan, 2001.

Iraqi Freedom: US and British invasion of Iraq, 2003.

Operation Mongoose

Attempts on the life and reputation of Cuba's President Fidel Castro planned by US agencies

1. Hiring Mafia hit men to assassinate him.
2. Giving him a scuba diving outfit infected with tuberculosis.
3. Booby-trapping a seashell that would explode if he lifted it from the seabed while diving.
4. Contaminating his favourite brand of cigars with the untraceable botulinum toxin.
5. Persuading him to write with a poisonous fountain pen.
6. Spiking his drinks with poison.
7. Shooting him with a sniper rifle.

8. Putting a chemical in his shoes that would make his beard fall out.
9. Operation Dirty Trick: A plot to blame Castro if the 1962 Mercury space flight carrying John Glenn crashed.
10. Operation Good Times: Faked photos of an obese Castro with two voluptuous women in a lavishly furnished room and a table laden with fine food. The caption would read, 'My ration is different'.

Items smuggled to Allied Prisoners in Colditz Castle by MI9 – Britain's escape and evasion service

1. Railway timetables
2. Information on sentries and frontiers
3. Food ration stamps
4. German currency
5. Maps
6. Compasses
7. Fake identity papers
8. An architect's blueprint of the castle

Methods Used to Attempt Escape from Colditz

1. Walking out of the main gates dressed in German uniform.
2. Dressing in German uniform and 'relieving' the sentries on duty.
3. Dressing up as a German housewife.
4. Running off during the daily stroll in the park.
5. Replacing prisoners with dummies while they hid under leaves in the park.
6. Vaulting over the wall.
7. Abseiling from the windows on ropes made from bed sheets.
8. Bribing the guards.
9. Tunnelling under the canteen.
10. Hiding inside surplus mattresses sent back to the town.
11. Sliding down the laundry chute.

12. Hiding in the dustcart.
13. Breaking through a lavatory wall.
14. Climbing through a manhole in the park.
15. Digging a tunnel under the chapel.
16. Hiding in the cart used to transport the dug-out earth to the town after the tunnel was discovered.
17. Manufacturing a full-size glider that could be launched from the castle roof carrying two passengers. The castle was liberated before it could be used, but a replica built later worked perfectly.

Bad predictions

Four or five frigates will do the business without any military force.

LORD NORTH ON THE AMERICAN REVOLUTION, 1774

No militia will ever acquire the habits necessary to resist a regular force.

GEORGE WASHINGTON, 1780

The Cavalry will never be scrapped to make room for the tanks; in the course of time Cavalry may be reduced as the supply of horses in this country diminishes. This depends greatly on the life of fox hunting.

JOURNAL OF THE UNITED SERVICES INSTITUTE, 1921

Some enthusiasts today talk about the probability of the horse becoming extinct and prophesy that the aeroplane, the tank and the motor-car will supersede the horse in future wars… I am sure that as time goes on you will find just as much use for the horse – the well-bred horse – as you have done in the past.

SIR DOUGLAS HAIG, 1925

People have been talking about a 3,000-mile high-angle rocket shot from one continent to another carrying an atomic bomb, and so directed as to be a precise weapon which would land on a certain target such as this city. I say technically I don't think anybody in the world knows how to do such a thing, and I feel confident it will not be done for a very long period of time to come. I think we can leave that out of our thinking.

VANNEVAR BUSH,
US CHIEF GOVERNMENT SCIENTIST, 1945

You will not need your rifles – all the Germans will be dead in their trenches.

BRITISH OFFICER AFTER THE ARTILLERY BARRAGE
THAT PRECEDED THE BATTLE OF THE SOMME, 1916

My good friends, for the second time in our history, a British Prime Minister has returned from Germany bringing peace with honour. I believe it is 'peace for our time.' Go home and get a nice quiet sleep.

NEVILLE CHAMBERLAIN, 30 SEPTEMBER 1938, ON HIS
RETURN FROM THE MUNICH CONFERENCE

In three weeks England will have her neck wrung like a chicken.

MARSHAL PÉTAIN, 1940 (PROMPTING CHURCHILL'S LATER
REMARK 'SOME CHICKEN! SOME NECK!')

Aeroplanes are interesting toys but of no military value.

GENERAL (LATER MARSHAL AND SUPREME COMMANDER OF
ALLIED FORCES IN 1918) FERDINAND FOCH, 1911

To throw bombs from an airplane will do as much damage as throwing bags of flour. It will be my pleasure to stand on the bridge of any ship while it is attacked by airplanes.

NEWTON BAKER, US MINISTER OF DEFENSE, 1921

Even if a submarine should work by a miracle, it will never be used. No country in this world would ever use such a vicious and petty form of warfare.

WILLIAM HENDERSON, BRITISH ADMIRAL, 1914

Hard pressed on my right; my left is in retreat. My centre is yielding. Impossible to manoeuvre. Situation excellent. I am attacking.

GENERAL FERDINAND FOCH TO GENERAL JOFFRE
DURING THE BATTLE OF THE MARNE, 1914

You're planning to make a ship sail against wind and tide by lighting a fire below deck? I don't have time to listen to that kind of nonsense.

NAPOLEON BONAPARTE ON ROBERT FULTON'S
PLANS TO MAKE A STEAMBOAT

We should declare war on North Vietnam . . . We could pave the whole country and put parking strips on it, and still be home by Christmas.

RONALD REAGAN, 1965

Letter allegedly written by the Duke of Wellington to the British Foreign Office in London from Central Spain, August 1812

Gentlemen,

Whilst marching from Portugal to a position which commands the approach to Madrid and the French forces, my officers have been diligently complying with your requests which have been sent by H.M. ship from London to Lisbon and thence by dispatch to our headquarters.

We have enumerated our saddles, bridles, tents and tent poles, and all manner of sundry items for which His Majesty's Government holds me accountable. I have dispatched reports on the character, wit, and spleen of every officer. Each item and every farthing has been accounted for, with two regrettable exceptions for which I beg your indulgence.

Unfortunately the sum of one shilling and ninepence remains unaccounted for in one infantry battalion's petty cash and there has been a hideous confusion as to the number of jars of raspberry jam issued to one cavalry regiment during a sandstorm in western Spain. This reprehensible carelessness may be related to the pressure of circumstance, since we are at war with France, a fact which may come as something of a surprise to you gentlemen in Whitehall.

This brings me to my present purpose, which is to request elucidation of my instructions from His Majesty's Government so that I may better understand why I am dragging an army over these barren plains. I construe that perforce it must be one of two alternative duties, as given below. I shall pursue either one with the best of my ability, but I cannot do both:

1. To train an army of uniformed British clerks in Spain for the benefit of the accountants and copy-boys in London or perchance,
2. To see to it that the forces of Napoleon are driven out of Spain.

Your most obedient servant,
Wellington

Foreign Terms and Phrases

Al-Qa'ida	'The Foundation' (Arabic)
Asibiya	'The companionship of warriors' (Arabic)
Blitzkrieg	Lightning war (German)
Bushido	'The way of the warrior' (Japanese)
Esprit de corps	'Team spirit' on the battlefield (French)
Fedayeen	'Those who sacrifice themselves' (Arabic)
Guerre couverte	Term for a war involving minor noblemen of the Middle Ages who had the authority to kill and maim but not to take prisoners or damage property (French)
Guerre de course	French term for naval warfare directed specifically against a nation's seaborne trade
Intifada	'Shaking off', referring to the Arab uprising in

	the West Bank and Gaza Strip in December 1987 (Arabic)
Mujahidin	'Warriors of God' (Arabic)
Samurai	'One who serves' (from the Japanese)
Sinn Fein	'Ourselves Alone' (Gaelic)
Stalin	'Man of Steel' (Russian)

The Balfour Declaration

'His Majesty's Government view with favour the establishment in Palestine of a national home for the Jewish people, and will use their best endeavours to facilitate the achievement of this object, it being clearly understood that nothing shall be done which may prejudice the civil and religious rights of existing non-Jewish communities in Palestine, or the rights and political status enjoyed by Jews in any other country.'

BRITISH FOREIGN SECRETARY ARTHUR JAMES BALFOUR,
2 NOVEMBER 1917

Arab–Israeli Wars

1. Israel's War of Independence, 1948–49
2. Sinai War, 1956
3. Six-Day War, 1967
4. War of Attrition, 1969–70
5. Yom Kippur War, 1973
6. War in Lebanon, 1982

Nuclear Stockpiles 2003

1. USA – 7,206 warheads
2. Russian Federation – 5,972
3. France – 464
4. China – 290
5. UK – 185
6. Israel – 150
7. India – 60
8. Pakistan – 24
9. North Korea – 3

Medieval Arms and Equipment

Arbalest	A hand crossbow
Aventail	A medieval hood of mail suspended from a basinet to protect the neck and shoulders
Barding	Armour for horses
Bastard sword	A longsword that could be wielded one or two handed
Basinet	An open-faced helmet
Bracer	A leather wrist-guard used by archers
Buckler	A small, round shield
Caltrop	Metal spikes that were placed on the ground to injure horses' hoofs
Caparison	Padded cloth or leather covering for a warhorse
Coif	A chain mail hood
Culverin	The smallest variety of cannon
Cuirass	Plate armour for the chest and back
Destrier	A knight's warhorse
Falchion	A curved short sword used by archers
Falcon	Medium cannon
Glaive	A pole arm with curved blade
Greaves	Shin guards
Halberd	A pole arm with an axe blade on one side of the head and a sharp spike on the other
Hauberk	A chain mail shirt with long sleeves
Heater	A shield with a straight top side and two curved sides meeting in a point at the bottom
Jack	Quilted fabric used as armour by English archers
Misericord	A single-edged hiltless dagger used for giving the *coup de grâce* to the seriously injured
Pennon	A triangular flag carried on the end of a knight's lance
Target	A small, round shield

Scorched Earth Policies

1. SHERMAN'S MARCH TO THE SEA: After reducing much of Atlanta to ruins in 1864, General Sherman began marching his 62,000 troops 250 miles to the coast across a 60-mile wide front. He promised to 'make Georgia howl' and burned or demolished virtually every bridge, railroad, factory, warehouse and barn in his path.

2. CHINA: In July 1940 Japanese General Yasuji Okamura initiated a strategy he called 'Take all, burn all, kill all' that reduced the Chinese population by several million in just eighteen months.

3. UKRAINE: In the face of the German advance in 1941, Soviet forces moved 6 million cattle, 550 large factories, thousands of small factories, 300,000 tractors, the country's entire railway rolling stock and 3.5 million skilled workers to the Urals while destroying most of the infrastructure and resources that remained. During their own retreat in 1943–44, the Germans razed 28,000 villages and 714 cities and towns, leaving 10,000,000 people homeless.

4. KUWAIT: In their retreat from Kuwait in 1991, Iraqi forces sabotaged over 700 of the country's oil wells, of which more than 600 were set on fire.

5. AFGHANISTAN: In 1999, the Taliban looted and burned up to 300 houses a day in the north of the country to prevent fleeing refugees from returning.

6. EAST TIMOR: Before the Indonesian army ended its occupation of East Timor in 1999, it systematically burned most of the country's towns and villages, damaging or destroying up to 75 per cent of Timorese buildings.

'The Feast of Crispian'

From *Henry V* by William Shakespeare, Act IV, Scene III

This day is call'd the feast of Crispian:
He that outlives this day, and comes safe home,
Will stand a tip-toe when this day is nam'd,
And rouse him at the name of Crispian.
He that shall live this day, and see old age,
Will yearly on the vigil feast his neighbours,
And say, 'To-morrow is Saint Crispian':
Then will he strip his sleeve and show his scars,
And say, 'These wounds I had on Crispin's day'.
Old men forget: yet all shall be forgot,
But he'll remember with advantages
What feats he did that day. Then shall our names,
Familiar in his mouth as household words,
Harry the king, Bedford and Exeter,
Warwick and Talbot, Salisbury and Gloucester,
Be in their flowing cups freshly remember'd.
This story shall the good man teach his son;
And Crispin Crispian shall ne'er go by,
From this day to the ending of the world,
But we in it shall be remembered;
We few, we happy few, we band of brothers;
For he to-day that sheds his blood with me
Shall be my brother; be he ne'er so vile
This day shall gentle his condition:
And gentlemen in England, now a-bed
Shall think themselves accurs'd they were not here,
And hold their manhoods cheap whiles any speaks
That fought with us upon Saint Crispin's day.

African Americans in the US Army

War of Independence – 5,000
Civil War – 200,000
First World War – 367,000
Second World War – 1,000,000
Korean War – 3,100
Vietnam War – 274,937
Persian Gulf War – 104,000

US War Dead

War of Independence (1775–83): 217,000 served – 4,435 killed
in battle
War of 1812 (1812–15): 286,730 served – 2,260 killed
Indian Wars (1817–98): approx. 106,000 served – approx. 1,000
killed
Mexican War (1846–48): 78,789 – 1,733 killed
Civil War (1861–66): 3,263,363 served – 214,938 killed
Spanish-American War (1898): 307,420 served – 385 killed
First World War (1917–18): 4,743,826 served – 53,513 killed
Second World War (1941–45): 16,353,659 served – 292,131 killed
Korean War (1950–53): 5,764,143 served – 33,667 killed
Vietnam War (1964–73): 8,752,000 served – 47,393 killed
Persian Gulf War (1991): 467,939 served – 148 killed

A US Serviceman's Chances of Death in Battle

War of Independence: 2 per cent (a 1 in 50 chance)
War of 1812: 0.8 per cent (1 in 127)
Indian Wars : 0.9 per cent (1 in 106)
Mexican War: 2.2 per cent (1 in 45)
Civil War: 6.6 per cent (1 in 15)
Spanish-American War: 0.1 per cent (1 in 798)
First World War: 1.1 per cent (1 in 89)
Second World War: 1.8 per cent (1 in 56)

Korean War: 0.6 per cent (1 in 171)
Vietnam War: 0.5 per cent (1 in 185)
Persian Gulf War: 0.03 per cent (1 in 3,162)

*Other chances of death for an average American over his or her lifetime:

Accident (land, sea, air) – 1.3 per cent (1 in 77)
Suicide – 0.8 per cent (1 in 122)
Murder – 0.5 per cent (1 in 211)
Narcotics – 0.2 per cent (1 in 592)
Fishing – 0.14 per cent (1 in 714)

British War Dead since the Second World War

Indonesia (1945–46)	50	killed
Palestine (1945–48)	223	
Malaya (1948–61)	509	
Korea (1950–53)	865	
Egypt (1951–54)	53	
Kenya (1952)	12	
Cyprus (1955–58)	79	
Egypt (1956)	12	
Borneo and Malaya (1962–66)	59	
Radfan (1964–67)	24	
Aden (1964–67)	68	
Falklands (1982)	250	
Persian Gulf War (1991)	42	

Some Notable War Films

All Quiet on the Western Front, dir. Lewis Milestone, 1930*
La Grand Illusion, dir. Jean Renoir, 1937
In Which We Serve, dir. Noel Coward and David Lean, 1942
The Cruel Sea, dir. Charles Frend, 1953*
Stalag 17, dir. Billy Wilder, 1953
The Bridge on the River Kwai, dir. David Lean, 1957*
Paths of Glory, dir. Stanley Kubrick, 1957

The Guns of Navarone, dir. J. Lee Thompson, 1961*
Lawrence of Arabia, dir. David Lean, 1962
The Longest Day, dir. K. Annakin, A. Marton, B. Wicki, D. F. Zanuck, 1962*
The Great Escape, dir. John Sturges, 1963*
Zulu, dir. Cy Endfield, 1964
Dr Strangelove, dir. Stanley Kubrick, 1964
Patton, dir. Franklin J. Shaffner, 1970
MASH, dir. Robert Altman, 1970
Cross of Iron, dir. Sam Peckinpah, 1977
Apocalypse Now, dir. Francis Ford Coppola, 1979
The Deer Hunter, dir. Michael Cimino, 1978
Das Boot, dir. Wolfgang Petersen, 1981 *
Full Metal Jacket, dir. Stanley Kubrick, 1987
Saving Private Ryan, dir. Steven Spielberg, 1998
Three Kings, dir. David O. Russell, 1999

* Adapted from books

Some Notable War Novels

War and Peace, by Leo Tolstoy, 1894
The Red Badge of Courage, by Stephen Crane, 1895
All Quiet on the Western Front, by Erich Maria Remarque, 1929
A Farewell to Arms, by Ernest Hemingway, 1929
For Whom the Bell Tolls, by Ernest Hemingway, 1940
The Naked and the Dead, by Norman Mailer, 1948
From Here to Eternity, by James Jones, 1951
Wheels of Terror, by Sven Hassel, 1959
Catch-22, by Joseph Heller, 1961
Slaughterhouse Five, by Kurt Vonnegut, 1969
Regeneration, by Pat Barker, 1991
Birdsong, by Sebastian Faulks, 1993

Great Sieges

1. *Azotus (now Ashdod) in Israel*, 664–610 BC: Besieged by Egyptian forces under Psamtik I for twenty-nine years

2. *Siege of Athens*, 404 BC: Brought the Athenian Empire to an end, with the demolition of the city walls and the Spartan Lysander's imposition of the Thirty Tyrants' oligarchy

3. *Carthage*, 149–146 BC: The end of the siege marked the death of the Carthaginian civilization. The inhabitants of the capital were either killed or enslaved and its buildings razed to the ground.

4. *Masada*, AD 73–74: The Judaean fortress of Masada was the last to be taken by the Romans as they reasserted imperial rule after the Jewish Rebellion. The 960 defenders committed mass suicide rather than surrender.

5. *Candia* (now Heraklion, Crete), 1648–69: The Turks besieged the Venetians for twenty-one years before their victory in 1669.

6. *Constantinople* (now Istanbul), 1453: Mehmed II of the Turks used various tactics, including concentrated cannon fire to breach the walls and transporting his navy 10 miles overland to attack the harbour, before a carelessly unlocked gate allowed the invaders into the city.

7. *Fort Sumter*, 1863–65: Confederate forces endured one of the longest sieges in modern warfare during their occupancy of the stronghold. For over 587 days from August 1863, 46,000 shells, estimated at over 7 million pounds of metal, were fired at the defenders. The three-storey structure was eventually reduced to rubble.

8. *Leningrad*, 1941–44: During an 880-day siege by the German Army from 30 August 1941 until 27 January 1944, an estimated 1.4 million defenders and citizens died. 641,000 starved to death, while 17,000 died in artillery strikes. Over 150,000 shells and 100,000 bombs fell on the city.

9. *Stalingrad*, 1943: The total death toll for the siege was 2,100,000. At the lifting of the siege, the civilian population stood at 1,515 from a pre-war count of over 500,000.

10. *Sarajevo*, 2 May 1992 to 26 February 1996: The Yugoslav National Army besieged the capital of Bosnia-Herzegovina for a total of 1,395 days.

'The Battle of Britain is About to Begin'

Excerpt from a speech given by Sir Winston Churchill to the House of Commons on 18 June 1940

'What General Weygand called the Battle of France is over. I expect that the Battle of Britain is about to begin. Upon this battle depends the survival of Christian civilization. Upon it depends our own British life, and the long continuity of our nation is turned on us. Hitler knows that he will have to break us in this Island or lose the war. If we can stand up to him, all Europe may be free and the life of the world may move forward into broad, sunlit uplands. But if we fail, then the whole world, including the United States, including all that we have known and cared for, will sink into the abyss of a new Dark Age made more sinister, and perhaps more protracted, by the light of perverted science. Let us therefore brace ourselves to our duties, and so bear ourselves that, if the British Empire and Commonwealth last for a thousand years, men will still say, This was their finest hour.'

Six Confederate Units in the American Civil War

Tallapoosa Thrashers
Bartow Yankee Killers
Chickasaw Desperadoes
Lexington Wildcats
Raccoon Roughs
South Florida Bulldogs

Military Artisans, Engineers and Designers

1. ARCHIMEDES: When the Romans under Marcus Claudius Marcellus laid siege to Syracuse in 213 BC, the city's most famous inhabitant designed a range of stone- and dart-throwing machines that took a heavy toll on the attackers. Particularly feared was the 'Claw' – a device lowered from the coastal walls that could capsize galleys.

2. CALLINICUS: The seventh-century Syrian engineer who invented Greek fire, the secret weapon of the Eastern Roman Emperors. The 'liquid fire' was thrown by siphons on to ships, where it burst into flames on contact and was virtually inextinguishable, even on water.

3. MASTER URBAN: The Hungarian cannon-maker could not find employment in the Byzantine Empire, so he sold his services to the Turks, who used his giant weapons to breach the walls of Constantinople in 1452. Urban was killed during the siege when one of his creations exploded.

4. SAMUEL COLT: Patented the Colt revolver in 1836, the handgun that 'made men equal'

5. DR RICHARD JORDAN GATLING: Invented the world's first machine gun in 1862, a hand-cranked weapon at first rejected by the Civil War forces of the time. Gatling initially trained as a medical doctor, and came to firearms through designing farm machinery.

6. HIRAM MAXIM: The inventor of the Maxim gun was at the Paris Electrical Exhibition in 1881 when he was told: 'If you want to make a lot of money, invent something that will enable these Europeans to cut each other's throats with greater facility.'

7. ANTHONY FOKKER: The Dutchman credited with devising a mechanical method of synchronizing machine-gun fire through the arc of an aircraft's propeller in 1914 – a system used to great effect by the Germans on Fokker's monoplanes.

8. SIR ROBERT WATSON-WATT: Credited with the production of a reliable RADAR system in 1935.

9. R.J. MITCHELL: Designer of the Spitfire fighter plane who died of cancer in 1937 before he could see the use to which it was put. Mitchell described the machine's title as 'a bloody silly sort of name', and the Spitfire was almost named the 'Shrew'.

10. BARNES WALLACE: Designed the bouncing bombs used by 617 Squadron in the Dambusters raid to crack the Ruhr dams in 1943.

11. WERNHER VON BRAUN: Leader of the German team of scientists that designed the V2 ballistic missile in 1942 and who later helped build the Saturn V rocket that took America to the moon.

12. J. ROBERT OPPENHEIMER: American director of the Manhattan Project and builder of the first atomic bomb.

13. MIKHAIL KALASHNIKOV: Russian designer of the ubiquitous AK-47 assault rifle, of which more than 70 million have been produced worldwide since 1949.

14. J. MIKE O'DWYER: Australian inventor of 'Metal Storm', the next generation machine gun with no moving parts, that fires up to a million rounds a minute electronically.

Found in a Pill Box at Passchendaele in 1917 after its recapture by Allied troops

SPECIAL ORDERS TO NO.1 SECTION

1. This position will be held and the section will remain here until relieved.
2. The enemy cannot be allowed to interfere with this programme.
3. If this section cannot remain here alive it will remain here dead but in any case it will remain here.
4. Should any man through shell-shock or such cause attempt to surrender he will stay here dead.
5. Should all guns be blown out the section will use Mills grenades and other novelties.
6. Finally the position as stated will be held.

CAMPBELL CPL

Battle Tanks Compared

Machine Country Entered Service	Weight	Armour	Main gun	Speed
FIRST WORLD WAR				
Mk 1 Britain 1916	62,720lb 28,450kg	12mm	2 × 6 pounder guns, 4 × 8mm machine guns, or 4 × 7.7mm machine guns, one 8 mm machine gun	3.7mph 5.95kmh
Renault FT-17 France 1918	15,432lb 7,000kg	22mm	8mm machine gun	4.7mph 7.7kmh
A7V Germany 1918	65,918lb 29,900kg	30mm	57mm	5mph 8kmh
SECOND WORLD WAR				
Char B1 France 1936	70,548lb 32,000kg	60mm	75mm	17mph 28 kmh
PzKpfw IV (Panzer Mk IV) Germany 1936	43,431lb 19,700kg	80mm	75mm	25mph 40kmh
Type 97 Chi-Ha Japan 1938	33,069lb 15,000kg	25mm	57mm	24mph 38kmh
A12 Matilda II Britain 1939	59,360lb 26,926kg	78mm	2 pounder	15mph 24kmh
T-34/85 Soviet Union 1940	70,547lb 32,000kg	60mm	85mm	31mph 50kmh
Carro Armato M13/40 Italy 1940	30,865lb 14,000kg	42mm	47mm	20mph 32kmh
M4 Sherman USA 1942	69,565lb 31,304kg	76mm	75mm	26mph 42kmh

Machine Country Entered Service	Weight	Armour	Main gun	Speed
PzKpfw VI Tiger I Germany 1942	121,253lb 55,000kg	100mm	88mm	24mph 38kmh
IS-2 (Josef Stalin) Soviet Union 1943	101,963lb 46,250kg	132mm	122mm	23mph 37kmh
PzKpfw V Panther Germany 1943	98,766lb 44,800kg	100mm	75mm	29mph 46kmh

COLD WAR AND AFTER

M60 USA 1960	108,000lb 48,600kg	120mm	105mm	30mph 48kmh
T-62 Soviet Union 1963	80,468lb 36,500kg	170mm	115mm	31mph 50kmh
Leopard I West Germany 1965	88,185lb 40,000kg	70mm	105mm	40mph 65kmh
AMX-30 France 1967	79,366lb 36,000kg	50mm	105mm	40mph 65 kmh
Challenger II Britain 1991	137,789lb 62,500kg	(300mm)*	120mm	36mph 59kph
M1 A2 Abrams USA 1998	151,872lb 68,888kg	(300mm)*	120mm	42mph 68kph

* equivalent to 300mm conventional armour in terms of strength and effectiveness

Poets Corner

Poets and writers of the First World War commemorated in Poet's Corner, Westminster Abbey, London.

Richard Aldington	David Jones
Laurence Binyon	Robert Nichols
Edmund Blunden	Wilfred Owen*
Rupert Brooke*	Herbert Read
Wilfrid Gibson	Isaac Rosenberg*
Robert Graves	Siegfried Sassoon
Julian Grenfell*	Charles Sorley*
Ivor Gurney	Edward Thomas*

* Died during the war

Notable War Correspondents and Photographers

1. **William Howard Russell:** *The Times*'s man in Crimea covered the charge of the Light Brigade and exposed both the suffering of the ordinary soldiers and the incompetence of their leaders.
2. **Archibald Forbes:** Witnessed Napoleon III's surrender to Bismarck in a weaver's cottage during the Franco-Prussian War and in 1879 once rode through 10 miles of Zulu territory to a telegraph office where he could report on the Battle of Ulundi.
3. **Januarius Aloysius MacGahan:** The crusading Irish American newsman revealed the massacre of 12,000 Bulgarian men, women and children at the hands of Kurdish forces commanded by Turkey in 1876.
4. **Stephen Crane:** The author of *The Red Badge of Courage* covered the charge up San Juan Hill by the Rough Riders and personally received the surrender of the town of Juana Diaz during the Spanish–American War of 1898.
5. **Cora Crane:** Stephen Crane's wife Cora, formerly the madam of an elegant brothel, was billed as the first female war correspondent. Writing for the New York press under the pen name Imogene Carter, she went with her husband to cover the Greco-Turkish war. After he died, she unsuccessfully attempted a

literary career of her own, then opened her second brothel in Jacksonville, Florida.

6. **Luigi Barzini:** The Italian reporter wrote of the brutality with which European troops put down the Boxer Rising in China, and told of the birth of the modern battlefield during the Russo-Japanese War of 1904.

7. **Winston Churchill:** Reported from, as well as participated in, the Battle of Omdurman and the Boer War.

8. **Robert Capa:** The photographer most famous for his coverage of the Spanish Civil War and the D-Day landings.

9. **Martha Gellhorn:** The veteran reporter who covered conflicts from the Spanish Civil War to the US invasion of Panama when she was 81.

10. **Alan Moorehead:** Chronicled the North African campaign of the Second World War in his classic *African Trilogy*.

11. **Dith Pran:** A Cambodian who remained in Phnom Penh to report on the fall of the city in 1975 and was subsequently interned in a Khmer Rouge re-education camp. He later escaped to the USA.

12. **Michael Herr:** An American journalist who wrote of drug-addled grunts in jungle combat in *Dispatches*, his account of the Vietnam War.

13. **Don McCullin:** British photographer most famous for his unflinching coverage of the Vietnam War.

14. **Jon Pilger:** An Australian broadcaster and war correspondent who has reported from many countries including Vietnam and Cambodia.

Notable Bridges and Bridgeheads

1. *Xerxes' Pontoon:* In 480 BC, Xerxes of Persia and his armies crossed the Hellespont, the strait separating Europe from Asia, on a floating bridge of 674 boats, tied together to make two parallel bridges, each with a length of 1.4km.

2. *Caesar and the Rhine:* In 55 BC Julius Caesar constructed a bridge across the upper Rhine in ten days. The German tribes

on the far bank were so in awe of this feat of engineering that they submitted to Roman power. Caesar dismantled the bridge and returned home having won without fighting a battle.

3. *The Potomac Pontoon*: In 1864, Union Army engineers constructed a 2,170-foot pontoon bridge of 68 boats linked by planks across the James River to enable Grant's Army of the Potomac, comprising some 45,000 men and 30,000 horses, to attack Petersburg, Virginia and cut off the Confederate Army's supply line.

4. *The Bridge on the River Kwai*: 260 miles/415km long and built in sixteen months beginning in October 1942. The total labour force consisted of about 68,000 Allied POWs and 200,000 Asian labourers, of whom 18,000 and 78,000 died respectively.

5. *A Bridge Too Far*: In September 1944, British and Polish paratroopers were charged with taking the bridge at Arnhem and holding it for 48 hours until armoured support arrived. Lieutenant-Colonel John Frost's battalion was the only one that reached the bridge, which it defended for seven days though the promised support never arrived.

6. *Remagen Bridge*: In the spring of 1945 the US 9th Armored Division took the Ludendorff railway bridge at Remagen in a surprise attack. Brigadier General Hoge proceeded against orders and raced units across before the Germans could detonate charges placed on the bridge.

7. *The Sava River Crossing*: Twenty thousand vehicles including the US Army's 1st Armored Division crossed the Sava River between Croatia and Bosnia and Herzegovina on a pontoon bridge 620 metres long amid severe floods in 1995.

The Amazon Women

'Amazon' = 'no breast' in Ancient Greek

1. CAUCASIANS: A tribe of warrior women who lived between the Black and Caspian Seas along the Thermidon River in around 8000 BC. At the age of 8, each girl had her right breast seared with a hot iron so that no mammary glands would grow to impede the use of a bow. The procedure was also believed to extirpate the masculine tendencies that were thought to emanate from the right side.

2. LIBYANS: A pre-Homeric tribe that lived along the Atlas Mountains of Morocco and wore red leather armour in battle.

3. GAGANS: North African Amazons who, until the tribe converted to Christianity, routinely killed baby boys.

4. HAMITICS: A tribe that lived between the Nile and the Red Sea.

5. SAUROMATIANS: A tribe that lived along the Don River in Russia. According to Herodotus, these women were not permitted to bear children until they had killed three male enemies.

6. EURYPLE'S AMAZONS: The tribe that captured Babylon in 1760 BC.

7. THE AMAZON AMAZONS: A tribe that attacked Portuguese explorers in the 1500s along the Amazon River.

Battlefield Trophies

1. **Armour:** From the ancient world to the Middle Ages, one of the chief tasks of a fighting man's squire was to strip his defeated opponents of their valuable armour as they lay on the battlefield.

2. **Roman Eagles:** Two eagle standards were taken by the barbarian forces of Arminius after the defeat of Varus in AD 9. They were recovered by Germanicus in AD 15 or AD 16.

3. **Ears:** After the Battle of Leigniz in 1241 the Mongols collected nine sacks of severed ears from the defeated forces of the Teutonic Knights.

4. **Spurs:** In 1302 an army, sent by Philip IV of France to put down the rebellious Flemish towns led by Bruges, was comprehensively

defeated. The spurs taken from the fallen French knights formed so large a mound that the battle was named after them.

5. **Scalps**: Taken by Native American tribes and also by the ancient Scythians, who made them into napkins and decorative clothing.

6. **Shrunken heads**: The Jivaro tribesmen of Ecuador traditionally decapitated their enemies and shrank their heads to the size of an orange; these were then worn like medals.

7. **Lugers**: The German Army issue automatic pistol was the most prized trophy for Allied troops during the Second World War. However, many who acquired one were injured due to the difficulty of operating its safety mechanism.

8. **Bones**: Skulls, vertebrae and other bones from dead Japanese servicemen were taken as souvenirs by US troops in the Pacific theatre in the Second World War. One lieutenant's young girlfriend posed with a skull she dubbed 'Tojo' for Life magazine's 'photo of the week'.

9. **Genitalia**: Around 1300 BC, King Menephta's army returned to Egypt with 13,000 severed phalluses taken from the defeated Libyans. Details were inscribed on a monument at Karnak: Libyan generals 6; Libyans 6,539; Sirculians 222; Etruscans 542; Greeks 6,111.

Nicknames of the Napoleonic Age

Given to British Army units in the eighteenth and nineteenth centuries

The Mongrels: The 7th Division – a mixed unit containing few British regiments.

The Cheeses: When social requirements were lowered for new recruits to the Life Guards in 1788, the regiment's officers complained that they were 'no longer gentlemen but cheesemongers'.

The Trades' Union: Given to the 1st Dragoon Guards, who were used to quell worker's riots.

The Rusty Buckles: Earned by the 2nd Dragoon Guards in a shambolic parade in Ireland.

The Virgin Mary's Bodyguard: Given to the 2nd Dragoon Guards after George II sent them to aid Maria Theresa of Austria.

The Bird Catchers: After the 1st Dragoons and 2nd Dragoons captured an Eagle at Waterloo in 1815.

The Saucy Seventh: Bestowed on the 7th Hussars for their smart uniforms.

The Cherry Pickers: Men from the 11th Light Dragoons were picking cherries in an orchard in Spain when they were surprised and captured by the French in 1811.

The Ragged Brigade: Given to the less well-dressed 13th Light Dragoons.

The Emperor's Chambermaids: After the 14th Light Dragoons captured King Joseph's silver chamberpot at Vitoria in 1813.

The Coalers: After the enlisted men of the 1st Foot Guards were hired out as labourers to shovel coal in order to raise funds to refurbish the officers' mess.

The Resurrectionists: Due to the number of men of the 3rd Foot who survived their wounds at Albuera in 1810.

Paddy's Blackguards: Because the 18th Foot was an Irish regiment.

The Nanny Goats: Because the regimental mascot of the 23rd Foot was a goat.

The Havercake Lads: A corruption of 'Have a cake, lad' – from the practice of the 33rd Foot's recruiting NCOs of tempting would-be recruits with oatcakes.

The Little Fighting Fours: Because there were so many short soldiers in the 44th Foot.

The Steelbacks: Due to the 57th Foot's predilection for flogging. The name was also given to the 58th Foot after Wellington had some of them flogged for stealing beehives in 1813.

The Bloodsuckers: After the 63rd Foot was decimated by disease spread by parasitic insects in the Caribbean in 1808.

The Decalogue – The Ten Commandments of the Knightly Code of Chivalry

1. Thou shalt believe all that the Church teaches, and shalt observe all its directions.
2. Thou shalt defend the Church.
3. Thou shalt respect all weaknesses, and shalt constitute thyself the defender of them.
4. Thou shalt love the country in which thou wast born.
5. Thou shalt not recoil before thine enemy.
6. Thou shalt make war against the Infidel without cessation, and without mercy.
7. Thou shalt perform scrupulously thy feudal duties, if they be not contrary to the laws of God.
8. Thou shalt never lie, and shall remain faithful to thy pledged word.
9. Thou shalt be generous, and give largesse to everyone.
10. Thou shalt be everywhere and always the champion of the Right and the Good against Injustice and Evil.

Young Lions

PRECOCIOUS MILITARY COMMANDERS

1. STEPHEN OF CLOYES: In 1212 the 12-year-old French shepherd led an army of several thousand boys to retake the holy land in the Children's Crusade. Stephen expected the waters of the Mediterranean to part for his advance, but when they did not, merchants offered to transport his forces across the sea free of charge. Once the children reached Brindisi, many were sold as slaves to the Moors, while the rest starved to death.
2. GENERAL THE MARQUIS GILBERT MOTIER DE LAFAYETTE: The 20-year-old French 'Boy General' commanded a recon- naissance force for George Washington during the American Revolutionary War and once led a successful breakout when his 2,000 troops were encircled by an army eight times their size.

3. GALUSHA PENNYPACKER: In 1863 the 23-year-old George Armstrong Custer was the youngest general in the Union Army. However, in 1865 Galusha Pennypacker was promoted to brigadier general one month before his 21st birthday – a US Army record that has stood ever since.

4. JOEL IGLESIAS: The young Cuban communist was only 15 years old when he joined Fidel Castro's rebels in 1957 and was an army commander when the revolutionary forces entered Havana in 1959.

5. CAPTAIN VALENTINE STRASSER: In 1992 Sierra Leone's president Joseph Momoh fled the state house when a delegation of army officers led by Strasser arrived to protest over late wages. The 26-year-old captain and champion disco dancer seized the chance to make himself military dictator of the country, which he went on to rule for four years.

6. JOHNNY AND LUTHER HTOO: Twin boys who led the ethnic Keren 'God's Army' rebel force against Mayanmar's military junta. They claimed to be 14 years old when they surrendered in 2001, but they were only eight according to other sources. Their Christian fundamentalist troops believed the boys to be divinely inspired.

The Ten Costliest Battles of the US Civil War

1. GETTYSBURG, 1–3 July 1863
   51,112 casualties (23,049 Union and 28,063 Confederate)
2. CHICKAMAUGA, 19–20 September 1863
   34,624 (16,170 Union and 18,454 Confederate)
3. CHANCELLORSVILLE, 1–4 May 1863
   30,099 (17,278 Union and 12,821 Confederate)
4. SPOTSYLVANIA, 8–19 May 1864
   27,399 (18,399 Union and 9,000 Confederate)
5. ANTIETAM, 17 September 1862
   26,134 (12,410 Union and 13,724 Confederate)
6. THE WILDERNESS, 5–6 May 1864
   25,416 (17,666 Union and 7,750 Confederate)

7. SECOND MANASSAS, OR BULL RUN, 29–30 August 1862
   25,251 (16,054 Union and 9,197 Confederate)
8. MURFREESBORO, OR STONE'S RIVER, 31 December 1862
   24,645 (12,906 Union and 11,739 Confederate)
9. SHILOH, 6–7 April 1862
   23,741 (13,047 Union and 10,694 Confederate)
10. FORT DONELSON, 13–16 February, 1862
    19,455 (2,832 Union and 16,623 Confederate)

Fighting on the Beaches

From a speech by Winston Churchill on 4 June 1940

'We shall not flag or fail. We shall go on to the end. We shall fight in France. We shall fight on the seas and oceans. We shall fight with growing confidence and growing strength in the air. We shall defend our island, whatever the cost may be. We shall fight on the beaches. We shall fight on the landing grounds. We shall fight in the fields and in the streets. We shall fight in the hills. We shall never surrender, and even if, which I do not for a moment believe, this Island or a large part of it were subjugated and starving, then our Empire beyond the seas, armed and guarded by the British fleet, would carry on the struggle, until, in God's good time, the New World, with all its power and might, steps forth to the rescue and the liberation of the old.'

Cannon Ammunition in the Age of Sail

1. ROUND SHOT: Stone balls before the seventeenth century, iron afterwards. These were used against the wooden hulls of enemy ships.
2. CHAIN SHOT: Two small round shot linked by a length of chain. This was used to slash through the rigging and sails of opposing ships.
3. CANISTER SHOT: Several hundred musket balls packed into a tin canister. Used mainly on land.

4. SPHERICAL CASE SHOT, OR SHRAPNEL: A hollow round shell packed with musket balls and a bursting charge, which exploded over the heads of the enemy.

5. GRAPE SHOT: Mainly used at sea, and consisting of a canvas bag filled with nine golf ball-sized solid balls, usually aimed at rigging and spars.

America's Wars and Foreign Interventions

1. The American Revolution, 1775–83 v. Great Britain
2. Indian Wars, 1775–1890 v. Native Americans
3. Quasi-War, 1798–1800 v. France
4. Barbary Wars, 1800–1815 v. The Barbary States (Tripoli, Algiers and Morocco)
5. War of 1812, 1812–15 v. Great Britain
6. Mexican–American War, 1846–48 v. Mexico
7. Intervention in Hawaiian Revolution, 1893 v. Hawaiian Government
8. Spanish–American War, 1898 v. Spain
9. Intervention in Samoan Civil War, 1898–99 v. German-backed forces
10. US–Philippine War, 1899–1902 v. Philippines
11. Boxer Rebellion, 1900 v. China
12. Moro Wars, 1901–13 v. Moslem Filipinos
13. Intervention in Panamanian Revolution, 1903 v. Colombia
14. The Banana Wars, 1909–33 v. Central American rebels
15. Occupation of Vera Cruz, 1914 v. Mexico
16. Pershing's Raid into Mexico, 1916–17 v. Mexico
17. First World War, 1917–18 v. Germany
18. Intervention in Russian Civil War, 1919–21 v. Bolsheviks
19. Second World War, 1941–45 v. Germany, Japan, Italy
20. Korean War, 1950–53 v. North Korea and China
21. Vietnam War, 1956–75 v. North Vietnam and South Vietnamese rebels
22. Intervention in Lebanon, 1958 v. Anti-government rebels

23. Intervention in the Dominican Republic, 1965 v. Anti-government rebels
24. Bombing of Libya, 1981 and 1986 v. Colonel Gaddafi's regime
25. Intervention in Lebanon, 1982–84 v. Syria, terrorist groups
26. Invasion of Grenada, 1983 v. Cubans and Grenadine communists
27. The Tanker War, 1987–88 v. Iran
28. Invasion of Panama, 1989 v. General Manuel Noriega's regime
29. Persian Gulf War, 1991 v. Iraq
30. Intervention in Somalia, 1992–94 v. Somali militia groups
31. Intervention in Bosnia, 1994–95 v. Bosnian Serbs
32. Occupation of Haiti, 1994 v. Haitian regime
33. Bombing of Afghanistan and Sudan, 1998 v. Terrorist groups
34. Operation Desert Fox, 1998 v. Iraq
35. Kosovo War, 1999 v. Serbia
36. Afghanistan War, 2001 v. Taliban and Al-Qa'ida
37. Iraq War, 2003 v. Iraq

America's Civil Wars and Rebellions

1. SHAY'S REBELLION, 1786–87
   Rebels v. the state government of Massachusetts
2. THE WHISKEY REBELLION, 1794
   Tax revolt in Western Pennsylvania
3. FRIES'S REBELLION, 1799
   'The Hot Water War' – Tax revolt in Pennsylvania
4. SLAVE REBELLIONS, 1800–65
   African American slaves
5. 'BLEEDING KANSAS', 1855–60
   Pro-slavery v. anti-slavery Kansans
6. BROWN'S RAID ON HARPER'S FERRY, 1859
   Rebellion against slavery led by John Brown
7. UNITED STATES CIVIL WAR, 1861–65
   Union (Northern) States v. Confederate (Southern) States

Largest Battleships

Vessel class	Nation	Year	Weight (tons)	Dimensions (feet)	Crew	Main guns	Speed (knots)
Yamato	Japan	1940	62,315	868×121	2,500	9×18.1in	27
Iowa	USA	1943	48,110	887×108	1,921	9×16in	33
Vanguard	Britain	1944	44,500	814×108	1,893	8×15in	30
Hood	Britain	1918	42,450	860×104	1,480	8×15in	32
Bismarck	Germany	1939	41,700	824×118	2,100	8×15in	30
Vittorio Veneto	Italy	1937	40,517	779×108	1,850	9×15in	30
South Dakota	USA	1941	37,970	680×108	1,793	9×16in	28
King George V	Britain	1939	36,727	745×103	1,422	10×14in	29
Richelieu	France	1939	35,000	813×108	1,670	8×15in	30
Scharnhorst	Germany	1936	34,850	754×98	1,670	9×11in	32

The Oldest British Regiments

1. *Royal Militia of the Island of Jersey* (1337): Now an engineering Field Squadron in the Territorial Army, the militia can be traced to the Anglo-Saxon fyrd of the ninth century, although most of units were formed in the eighteenth century. A forty-year disbandment after 1946 technically disqualifies them from the title of oldest regiment.

2. *Honourable Artillery Company* (1537): Although formed in 1296, they received the Royal Charter only in 1537. Their record of service to the Crown is not considered unbroken, as they fought on the side of the Roundheads during the English Civil War.

3. *Royal Monmouthshire Royal Engineers* (Militia) (1539): Although founded two years after the Honourable Artillery Company, the RMonRE takes precedence due its unbroken loyalty to the Crown.

4. *The Buffs* (1572): Formed from London's urban militia to support the Protestants in Holland, where they remained until the outbreak of the Anglo-Dutch war in 1665, at which point they

were disbanded for refusing the oath of loyalty to the Dutch States General. They fled to England and reformed as 'The Holland Regiment' in the British Army. The unit is now part of The Princess of Wales's Royal Regiment.

5. *The Connaught Rangers/Scots Brigade* (1568–1922): Formed for Dutch service but took the 1665 oath of allegiance to the Dutch States General, declining to return home. They joined the British Army as the 94th Foot in 1794 after rebelling against taking orders in the Dutch language. In 1881 they merged with the Connaught Rangers in Ireland and were disbanded when Ireland gained its independence in 1922.

6. *The Royal Scots* (1633): Formed by Charles I to fight for France in the Thirty Years' War, the Scots are the oldest regiment in the Regular Army. During a good-natured argument over seniority with the French Picardie regiment in the seventeenth century, the Scots claimed to be descended from the Roman unit that guarded Jesus' tomb. Not to be bettered, the French replied that had they been on guard instead, Jesus' body would not have gone missing. Thus they gained the nickname 'Pontius Pilate's Bodyguard'.

7. *Coldstream Guards* (1650): A Parliamentary unit formed during the English Civil War that entered royal service under Charles II. They are named after General Monck's march from Coldstream to London in 1660 and boast the longest continuous service of any regular regiment in the British Army.

8. *Royal Horse Guards* (1650): Formed to fight for the Roundhead cause, they are the oldest cavalry regiment and later became the Blues and Royals.

9. *First Footguards* (1656) until after Waterloo, then Grenadier Guards: Created by Charles II as his bodyguard on his return to England from Holland in 1660.

10. *Life Guards* (1658): Formed from a Parliamentary unit and a troop raised for the exiled Charles II.

Marching Rates – Paces Per Minute

Napoleonic Grande Armée – 120
British Army (line infantry and corps) – 120
Chinese People's Liberation Army - 108
Ancient Roman Legion – 100
French Foreign Legion – 88
British Army under Wellington – 75

Accidents at Sea

Ships sunk or damaged without enemy action

1. The *Mary Rose*: In 1545, Henry VIII's overloaded flagship wheeled in the wind and sank as water rushed into the lower gun ports that the crew had omitted to close.
2. The *Wasa*: The 64-gun Swedish flagship sank on her maiden voyage in 1628, capsizing in the wind almost as soon as her sails were hoisted.
3. The *Kronan*: The Swedish Admiral Baron Lorentz Creutz's last words were 'In the name of Jesus, make sure that the cannon ports are closed and the cannon made fast, so that in turning we don't suffer the same fate as befell the *Wasa*'. The ports were not closed and his flagship sank in 1675 as her predecessor had about fifty years earlier.
4. HMS *Association*: In 1707 the British Mediterranean commander, Sir Cloudesley Shovell, was returning home late in the season when his squadron misjudged their longitude and were wrecked on the Isles of Scilly. This disaster led the government to offer sponsorship for research into a reliable method of determining longitude, resulting in John Harrison's chronometer.
5. HMS *Victoria*: In an exercise in the Bay of Tripoli in 1893, Admiral Sir George Tryon of the Royal Navy put his flagship on a collision course with HMS *Camperdown*. He refused to reverse his orders despite the warnings of his officers and was reported to have said 'It is all my fault' before he drowned with 358 other seamen.

6. **U-28:** During the First World War, the German submarine launched a close-range surface attack on the British cargo ship *Olive Branch*. When a shell from U-28's deck gun set off a consignment of ammunition aboard the ship, the explosion was enough to sink the attacking submarine.

7. **HMS *Trinidad*:** In 1941 the Royal Navy submarine fired a torpedo at a German destroyer only for the weapon to describe a curving course – returning to destroy the Trinidad's engine room and put her out of the war. The same fate was suffered by the American submarine USS *Tang*, which torpedoed itself in the Formosa Strait in 1944.

8. **K-141 *Kursk*:** The nuclear-powered Russian submarine sank with all hands in the Barents Sea in 2000 after an explosion in one of her torpedo tubes. Russian authorities have claimed this was due to a collision with a foreign vessel – an account dismissed by international investigators.

Quotations

Dulce et decorum est pro patria mori
(It is a sweet and seemly thing to die for one's country)

HORACE, *ODES*, III

I want you to remember that no son of a bitch ever won a war by dying for his country. He won it by making the other poor dumb bastard die for his country.

GENERAL GEORGE PATTON

O people, know that you have committed great sins and that the great ones among you have committed these sins. If you ask me what proof I have for these words, I say it is because I am the punishment of God. If you had not committed great sins, God would not have sent a punishment like me upon you.

**GENGHIS KHAN TO THE SURVIVORS OF HIS
SACK OF BUKHARA, 1220**

Veni, vidi, vici
(I came, I saw, I conquered)

**JULIUS CAESAR AFTER DEFEATING KING
PHARNACES II IN ASIA MINOR, 47 BC**

I would like to see the clause in Adam's will which excludes France from the division of the world.

FRANÇOIS I (1494–1547)

Going to war without France is like going deer hunting without your accordion.

NORMAN SCHWARTZKOPF

I know I have the body of a weak and feeble woman, but I have the heart and stomach of a king, and of a king of England too; and think foul scorn that Parma or Spain, or any prince of Europe, should dare to invade the borders of my realm.

QUEEN ELIZABETH I TO THE ARMY AT TILBURY AS
THE SPANISH ARMADA APPROACHED ENGLAND IN 1588

It takes a brave man not to be a hero in the Red Army.

JOSEPH STALIN

Pour encourager les autres
VOLTAIRE ON THE EXECUTION OF ADMIRAL BYNG IN 1757 FOR
HAVING RETREATED IN THE FACE OF THE ENEMY

Oh! he is mad, is he? Then I wish he would bite some of my other generals.

GEORGE II, REPLYING TO ADVISORS WHO TOLD
HIM THAT GENERAL JAMES WOLFE WAS INSANE

Qui desiderat pacem, praeparet bellum.
(Let him who desires peace prepare for war)

VEGETIUS, FOURTH CENTURY

Political power grows out of the barrel of a gun.

MAO TSE-TUNG

I hope to God I have fought my last battle... I am wretched even at the moment of victory, and I always say that next to a battle lost, the greatest misery is a battle gained.

THE DUKE OF WELLINGTON AFTER WATERLOO

Ours is composed of the scum of the earth – the mere scum of the earth. The British soldiers are fellows who have all enlisted for drink – that is the plain fact – they have all enlisted for drink.

WELLINGTON DESCRIBING THE BRITISH ARMY, 1811

It takes 15,000 casualties to train a major general.

FERDINAND FOCH

An army marches on its stomach.

NAPOLEON

The British Army should be a projectile to be fired by the British Navy.

SIR EDWARD GREY (1862–1933)
TO THE HOUSE OF COMMONS

C'est magnifique, mais ce n'est pas la guerre.
It is magnificent, but it is not war.

PIERRE BOSQUET AFTER WITNESSING THE CHARGE OF
THE LIGHT BRIGADE, 25 OCTOBER 1854

War is an ugly thing, but not the ugliest of things. The decayed and degraded state of moral and patriotic feeling, which thinks that nothing is worth war, is much worse. The person who has nothing for which he is willing to fight, nothing which is more important than his own personal safety, is a miserable creature and has no chance of being free unless made and kept so by the exertions of better men than himself.

JOHN STUART MILL, 1868

'Nuts!'
In response to the demand that he surrender his surrounded troops at Bastogne.

MAJOR GENERAL ANTHONY MCAULIFFE, US ARMY,
23 DECEMBER 1944

Oh, it's Tommy this an' Tommy that an' 'Tommy, go away',
But it's 'Thank you Mister Atkins' when the band begins to play.
Then it's Tommy this an' Tommy that, an' 'Tommy 'ows yer soul?'
But it's 'Thin red line of 'eroes' when the drums begin to roll.

RUDYARD KIPLING, *TOMMY*

Great part of the information obtained in war is contradictory, a still greater part is false, and by far the greatest part is of a doubtful character.

KARL VON CLAUSEWITZ, *ON WAR*, 1832

Don't talk to me about Naval tradition. It's nothing but rum, sodomy and the lash.

WINSTON CHURCHILL TO THE BOARD OF ADMIRALTY, 1939

Young man, you did a very fine thing to give up a most promising career to fight for your country. Mark you, had you not done so, it would have been despicable.

WINSTON CHURCHILL TO DAVID NIVEN

Peccavi.
(I have sinned)

GENERAL NAPIER'S ALLEGED DISPATCH UPON
CAPTURING THE INDIAN PROVINCE OF SINDH, 1843

I shall return.

DOUGLAS MacCARTHUR UPON LEAVING THE PHILIPPINES
AHEAD OF THE JAPANESE INVASION FORCE, 1942
(HE DID RETURN, IN 1944)

I thought he was talking about our mess bill.

ANONYMOUS RAF PILOT UPON HEARING CHURCHILL'S
SPEECH CONTAINING THE LINES 'NEVER IN THE FIELD
OF HUMAN CONFLICT HAS SO MUCH BEEN OWED
BY SO MANY TO SO FEW'

Overpaid, oversexed and over here.

POPULAR BRITISH DESCRIPTION OF
AMERICAN SERVICEMEN IN THE SECOND WORLD WAR

Men go into the Navy thinking they will enjoy it. They do enjoy it for about a year, at least the stupid ones do, riding back and forth quite dully on ships. The bright ones find that they don't like it in half a year, but there's always the thought of that pension if only they stay in… Gradually they become crazy. Crazier and crazier. Only the Navy has no way of distinguishing between the sane and the insane. Only about 5 per cent of the Royal Navy have the sea in their veins. They are the ones who become captains. Thereafter, they are segregated on their bridges. If they are not mad before this, they go mad then. And the maddest of these become admirals.

GEORGE BERNARD SHAW

He's either never been to Umm Qasar or he's never been to Southampton. There's no beer, no prostitutes and people are shooting at us. It's more like Portsmouth.

BRITISH TROOPER AFTER GEOFF HOON,
THE DEFENCE MINISTER, HAD COMPARED THE IRAQI PORT
TO THE BRITISH COASTAL TOWN

They couldn't hit an elephant at this dist…

LAST WORDS OF UNION GENERAL
JOHN SEDGEWICK, 1864

SOURCES

The material in this book was compiled from several hundred sources. The most up-to-date figures and most comprehensive information have been given wherever possible. This has meant that data and statistics from different authorities have had to be combined in tables and lists that therefore cannot be attributed to a single origin for research purposes. However, among the most useful reference books were the *CIA World Fact Book 2002*, the *SIPRI Yearbook 2002,* Cassell's History of Warfare series and the Jane's series of defence titles. The Imperial War Museum, London, has been an invaluable resource. Many conflicting versions of historical events and statistics exist, and in these cases the most widely agreed upon information has been favoured.

ACKNOWLEDGEMENTS

Special thanks go to Angus MacKinnon, Penny Gardiner and Gordon Corrigan for their work in editing this book, to Philip Lewis for the jacket and text design, FiSH for the typesetting, and to Toby Mundy and Bonnie Chiang for overseeing the project from start to finish.